ICMI Study Series Editors: A. G. Howson and J.-P. Kahane

The Influence of Computers and Informatics on Mathematics and its Teaching

Editorial Board

R. F. Churchhouse

B. Cornu

A. G. Howson

J.-P. Kahane

J. H. van Lint

F. Pluvinage

A. Ralston

M. Yamaguti

The right of the
University of Cambridge
to print and sell
all manner of books
was granted by
Henry VIII in 1534.
The University has printed
and published continuously
since 1584.

CAMBRIDGE UNIVERSITY PRESS

Cambridge

London New York New Rochelle

Melbourne Sydney

217143

510
I 43

Published by the Press Syndicate of the University of Cambridge
The Pitt Building, Trumpington Street, Cambridge CB2 1RP
32 East 57th Street, New York, NY 10022, USA
10, Stamford Road, Oakleigh, Melbourne 3166, Australia

First published 1986

Printed in Great Britain at the University Press, Cambridge

Library of Congress cataloging in publication data:

The influence of computers and informatics on mathematics and its
teaching.
Proceedings from a symposium held in Strasbourg, France in March 1985
and sponsored by the International Commission on Mathematical
Instruction.
1. Mathematics - Study and teaching - Congresses. 2. Mathematics -
Computer assisted instruction - Congresses. 3. Computers - Congresses
4. Electronic data processing - Mathematics - Congresses.
I. Howson, A.G. (Albert Geoffrey), 1931- . II. Kahane, Jean-Pierre.
III. International Commission on Mathematical Instruction.
QA11.A1I425   1986   510   85 25536

British Library cataloguing in publication data:

The influence of computers and informatics on mathematics and its
teaching. - (International Commission on Mathematical Instruction).
1. Mathematics - Study and Teaching (Higher) - Data Processing
I. Howson, A.G.   II. Kahane, J.-P.   III. Series
510'.7'8   QA20.C65

ISBN   0 521 32402 5   hardcover
ISBN   0 521 31189 6   paperback

CONTENTS

Foreword

The International Commission on Mathematical Instruction is planning a series of studies on topics of current interest within mathematical education. The first study was on the influence of computers and informatics on mathematics and its teaching at university and senior high school level. It was prepared by a Program Committee, consisting of R.F. Churchhouse (Cardiff), B. Cornu (Grenoble), A.P. Ershov (Novosibirsk), A.G. Howson (Soughampton), J.-P. Kahane (Orsay), J.H. van Lint (Eindhoven), F. Pluvinage (Strasbourg), A. Ralston (Buffalo), M. Yamaguti (Kyoto). A discussion document was sent to all national delegates of ICMI and printed in l'Enseignement Mathématique, 30 (1984). Contributions written in response to this paper formed the basis of discussions at a symposium held in Strasbourg in March, 1985.

The Proceedings begin with a general report which looks in particular at the three themes:
1. How do computers and informatics influence mathematical ideas, values and the advancement of mathematical science?
2. How can new curricula be designed to meet changing needs and possibilities?
3. How can the use of computers help the teaching of mathematics?

There then follows a selection of papers contributed to the Strasbourg symposium. The selection is our responsibility, and is based on the recommendations of the Editorial Board. In some cases a paper has been selected because of its originality, in others because it represents a theme treated in several contributions – in such cases our choice reflects our wish to give prominence to the theme, and not necessarily our support for the arguments advanced in a particular paper.

Other written contributions can be found in a volume of supporting papers, published by IREM, Université Louis Pasteur, Strasbourg, and described elsewhere in this book.

Neither the Proceedings nor the supporting papers put an end to the discussion. On the contrary, they are intended to provide a sound basis for further discussion and action. ICMI does not plan to continue with this study at a general level, but is ready to help organize further investigations on particular topics within this field. If readers wish to initiate such investigations, we invite them to make contact either with us or with their national delegate to ICMI.

Many people have contributed greatly to this study. Here we have only space to mention and to thank Francois Pluvinage, who was responsible for administrating the Strasbourg meeting and for editing the Supporting Papers, to those bodies which gave financial assistance, including DCRI, the French Mathematical Society, IBM, ICSU, the Royal Society, UNESCO and the University of Strasbourg, to all who submitted papers or attended the Strasbourg meeting, and to Mrs. June Kerry who has so carefully typed the major part of this volume.

A.G. Howson
J.-P. Kahane

THE INFLUENCE OF COMPUTERS AND INFORMATICS

ON MATHEMATICS AND ITS TEACHING

PART I

THE EFFECT ON MATHEMATICS

1.1 Introduction

Mathematical concepts have always depended on methods of
calculation and methods of writing. Decimal numeration, the writing of
symbols, the construction of tables of numerical values all preceded
modern ideas of real number and of function. Mathematicians calculated
integrals, and made use of the integration sign, long before the emergence of Riemann's or Lebesgue's concepts of the integral. In a similar
manner, one can expect the new methods of calculation and of writing
which computers and informatics offer to permit the emergence of new
mathematical concepts. But, already today, they are pointing to the
value of ideas and methods, old or new, which do not command a place in
contemporary "traditional" mathematics. And they permit and invite us
to take a new look at the most traditional ideas.

Let us consider different ideas of a real number. There is a point on
the line R, and this representation can be effective for promoting
the understanding of addition and multiplication. There is also an
accumulation point of fractions, for example, continued fractions giving the best approximation of a real by rationals. There is also a
non-terminating decimal expansion. There is also a number written in
floating-point notation. Experience with even a simple pocket calculator can help validate the last three aspects. The algorithm of continued fractions – which is only that of Euclid – is again becoming a
standard tool in many parts of mathematics. Complicated operations
(exponentiation, summation of series, iterations) will, with the
computer's aid, become easy. Yet even these simplified operations will
give rise to new mathematical problems: for example, summing terms in
two different orders (starting with the largest or starting from the
smallest) will not always produce the same numerical result (see, e.g.,
Churchhouse, 1980, 1985).

Again, consider the notion of function. Teaching distinguishes between,
on the one hand, elementary and special functions – that is, those
functions tabulated from the 17th to the 19th century – and, on the
other, the general concept of function introduced by Dirichlet in 1830.
Even today, to "solve" a differential equation is taken to mean reduc-
ing the solution to integrals, and if possible to elementary functions.
However, what is involved in functional equations is the effective
calculation and the qualitative study of solutions. The functions in

which one is interested therefore are calculable functions and no longer only those which are tabulated. The theories of approximation and of the superposition of functions - developed well before computers - are now validated. The field of elementary functions is extended, and functions of a non-elementary nature are introduced naturally through the discretisation of non-linear problems. Informatics, too, compels us to take a new look at the notion of a variable, and at the link between symbol and value. This link is strongly exploited in mathematics (for example, in the symbolism of the calculus). In informatics, the necessity of working out, of realising the values has presented this problem in a new way. The symbolism of functions is not entirely transferable, and the attributes of a variable are different in languages such as FORTRAN, LISP and PROLOG.

In the sections that follow we look at some aspects of how computers and informatics have already affected mathematics and mathematical research and present some thoughts on what future effects might be seen. We do not claim that our survey is comprehensive, especially so in the disciplines of applicable mathematics, but we hope that it provides some pointers. In any event information technology, in the widest sense, is advancing far too fast for any predictions to be of value for a period of more than a few years.

1.2 New and revived areas of mathematical research

Computers are not only providing a new tool in mathematical research and teaching, they are, at the same time, themselves the source of new areas of research. Not all of the research stimulated by the availability of computers is in new branches of mathematics, some is of ancient lineage, going back to the 19th or 18th century, but open now to attack with a weapon not available to Euler, Gauss, Jacobi, Ramanujan and others. Who can doubt, though, that these giants of the past would have exploited these new possibilities with enthusiasm had they been available? It is one of the unique features of mathematics that it is based upon a body of results that never loses its value. Fashions and interests may change, but the neglected subject of the last century, or even of the last millenium, may prove to be of new interest at any time when conditions are right for its re-emergence. So the corpus expands; nothing ever dies, though it may remain dormant for centuries. In the age of information technology we wish to emphasise this fact, for it underlies everything that follows.

One of the most famous examples of mathematical research being stimulated by the use of a computer is the soliton (solitary wave) solution of the Korteweg-de Vries equation by Zabusky and Kruskal (1965), which was initially suggested by numerical results. Continuing experimental investigations have indicated the existence of other, related, solutions and theoretical research has provided a substantial framework for investigating soliton solutions of several non-linear wave equations.

Another example will be found in Yamaguti's work (Supporting Papers) which may be summarised briefly by saying that he observed continuous, but nowhere-differentiable, functions via numerical experiments on dynamical systems defined iteratively whose solutions exhibit very chaotic behaviour. Particular cases produce the Weierstrass function and the Takagi function; the latter may be written

$$T(X) = \sum_{k=1}^{\infty} 2^{-k} \phi^{(k)}(X)$$

where $\phi(X) = 2x \ (0 \leqslant x < \tfrac{1}{2}), \ = 2(1-x) \ (\tfrac{1}{2} \leqslant x \leqslant 1)$

and has recently been used in teaching elementary analysis. Further research, in collaboration with Hata on a family of finite difference schemes led to Lebesgue's Singular Function.

Among long-established branches of Pure Mathematics where computers have had a major impact are Group Theory, Combinatorics and Number Theory. Many applications of computers in these areas have been published in proceedings of conferences (for example, Churchhouse and Herz (1968), Atkin and Birch (1971), Leech (1970)).

The applications are already too numerous to list in full or describe in detail but it is clear that the search for sporadic groups, the investigation of Burnside's problem, the study of rational points on elliptic curves, and the search for large primes would be quite impossible without computers. The factorisation of large integers is another example; although intrinsically it is not an exciting topic it has recently assumed considerable importance in relation to cryptography and public-key systems (Beker and Piper, 1982). Many of these applications have benefited considerably from the availability of program packages specifically designed as an aid for researchers in the field; the CAYLEY system for the study of finite simple groups is a well-known example; such systems relieve research workers of a great deal of drudgery. Another 'old' topic that has taken on a new lease of life is that of continued fractions, both as providing approximations to real numbers and, in analytical form, in numerical analysis.

The availability of colour graphics displays and packages has opened up exciting possibilities for research not only in geometry, modelling and fluid flow but in less obvious areas such as analysis. The study of the iteration of complex-valued functions has been transformed recently; the complex nature of Julia sets and their descendants is made beautifully apparent by the use of colour graphics, even though their mathematical nature remains largely unknown (see, for example, Section 1.5 below and West (Supporting Papers)).

It is clear to us that the computer is having, and will continue to have, a significant impact on the directions of mathematics research, on the way in which mathematicians carry out their research and that computers will not only be commonly used to arrive at conjectures but also to assist in finding proofs. In addition some important questions are raised (i) how should computers be used to assist mathematicians in communicating their discoveries and in keeping abreast of the research of others? and (ii), what are likely to be the intellectual and social consequences, so far as mathematics and mathematicians are concerned, of the widespread interest in, and use of computers?

1.3 Proof

In mathematics a "proof" is, strictly, a chain of deductions from the axioms; in practice, of course, a proof is accepted if it makes use of results which have themselves been deduced from the axioms, or from other results, etc., etc. It would be possible, but exceedingly tedious, to write out a proof of the theorem that every positive integer is the sum of the squares of four integers starting from the axioms of arithmetic, but few people would regard this as necessary and would accept various intermediate steps - an identity of Jacobi, or representation of integers by binary quadratic forms - as valid rungs on the ladder, since each of these steps is deducible from other results which are deducible ..... from the axioms.

Computers might be used in mathematical proofs; they might, initially, suggest what is true and, equally important, what is not, they might be used for computations which are required in a proof; they might be used - as in the proof of the 4-colour theorem (Appel and Haken, 1976) - to examine all of a finite set of cases, on which the truth of the theorem ultimately depends; they might even be programmed to find part of the proof by trying many possible combinations of known axioms, theorems or identities, though the "combinatorial explosion" makes such an approach infeasible except in very special cases.

As examples, computers have been used to suggest results in group theory, combinatorics, number theory, coding theory and to support the truth of conjectures such as the Riemann Hypothesis. For an early survey article see Churchhouse (1973). Among notable theorems which were initially conjectured on the basis of numerical evidence are the Prime Number Theorem (Gauss) and several important results of Ramanujan (1927) including the congruence properties of the partition function and of the function $\tau(n)$. On the other hand Lander and Parkin (1967) and a computer found that

$$27^5 + 84^5 + 110^5 + 133^5 = 144^5$$

and so <u>disproved</u> a conjecture of Euler that had stood for nearly 200 years.

Accuracy and reliability of the computations should not be an issue today. Where a result is sufficiently important or in doubt it can be checked by someone else on a different machine; this has been done on several occasions and if the result is confirmed and, assuming that the underlying mathematics is correct, the result can be accepted with considerable confidence, if not certainty. Computer-assisted proofs need not be any more suspect than purely human proofs; many false "proofs" - including the 4-colour theorem - have been published in the past; we do not believe that the computer will increase the number of false proofs, quite the contrary.

It is, of course, accepted that no amount of numerical evidence constitutes a proof of a theorem relating to an infinite set; the numerical evidence may be misleading even for a very large set of values of the variables involved. A well-known example from analytic number theory is Littlewood's proof (see Ingham, 1932) that despite all the numerical evidence then, and even now, available

$$\pi(x) - \int_2^x \frac{dt}{\ln t}$$

($\pi(x)$ indicating the number of primes less than or equal to x)

not only eventually changes sign, but does so infinitely often.

A criticism of computer-assisted proofs - such as the 4-colour theorem - is that they tend to rely on brute-force and give little insight into why the theorem is true. Unfortunately some results e.g. finding large primes or factoring large integers intrinsically require such methods, and whilst it may be true that a computer proof may bring little insight its very existence may inspire people to find more elegant, shorter, or illuminating proofs.

Taking a longer-term view, the availability of computer assistance may encourage mathematicians to a more precise syntax and to express more formally what is in their minds (de Bruijn, below). Such a development may, in turn, aid the teaching of the art of constructing proofs and so lead to the development of 'expert systems' to undertake at least some aspects of mathematical work (including all the routine algebraic manipulation, computation, etc.), in partial fulfilment of Leibniz's dream of a rational calculating device.

One final point: since every proposition that is provable has among its many proofs one of minimal length and since the proofs of any given length are (at most) finite in number there must be true theorems of mathematics that cannot be demonstrated by traditional discourse within the longest human lifetime. It would appear then that there are mathematical theorems that can only be proved with the aid of computers if we are unwilling to wait too long.

1.4 Experimentation in Mathematics
Certain branches of mathematics have always been open to experimentation but the arrival of computers means the scope for experimentation in mathematics has been greatly increased. In some of the sections above we have indicated cases where experiments have been used to provide data on which conjectures and, in some cases, theorems have been based. Euler, remarking on the necessity of observation in mathematics, said: "The problems of numbers that we know have usually

been discovered by observation, and discovered well before their validity has been confirmed by demonstration ....."

The sheer speed of computers means that calculations which would once have taken a lifetime can now be completed in hours, or even minutes. Add to this the fact that the results can often, if required, be presented in graphical form rather than as a list of numbers and we see that the interpretation of the experiments may be made much easier. The case of the iteration of complex valued functions illustrates this point.

Of course when a constraint is relaxed there is a danger of excess. The ability to perform calculations does not mean that everything can or should be calculated. There is a balance to be struck and this must be guided by experience - not to mention the cost of the computations. The effort and cost involved need to be combined with the probability of success, in the sense of solving a problem or uncovering some useful fact. Computation for the sake of computation is not to be encouraged.

Although experimentation in pure mathematics has its uses it is, perhaps, in the area of statistics that it is particularly valuable. We take two examples.

Simulation

Even before the availability of the modern computing technology, experimental sampling and Monte Carlo methods have played a role in statistics for studying the performance of statistical techniques under the assumption of probability models. The computer has enhanced this aspect on a large scale. One famous example is the Princeton Robustness Study (Andrews et al, 1972) where sets of estimators under a system of different modelling assumptions are studied by means of computer simulation. The results have stimulated new mathematical research into robust estimators (e.g. asymptotic theory) but on the other hand they cannot merely be interpreted as conjectures that can and should be validated by mathematical proof, but they have an importance in itself and have already influenced the practice of analysing data.

Exploratory Data Analysis

It is sometimes stated that the computer has led to an unwelcome shift from hard thinking to a senseless computation of examples and experimentation. A balanced picture would say that the computer has led to broader variety of "types of rationality" to approach problems and it is necessary to judge in every situation which approach is more reasonable.

The classical paradigm for applying statistics is to think first very hard then construct a probabilistic model and an adequate design for gathering data. But this strategy is not feasible in quite a lot of situations where little is known about the data and the underlying

system of interest. In connection with the numerical and graphical
capabilities of computers a new methodology of data analysis, called
Exploratory Data Analysis (Tukey, 1977), has been developed. The
computer has made it possible to experiment with several models for a
data set, to construct a variety of interesting plots of the data to
gain insights into patterns, structures and anomalies of the data and
to develop conjectures concerning the features of the system underlying
the data. Such a type of exploratory mathematics would not be practic-
able on a large scale without using computers.

1.5 Iterative methods
Methods of solving systems of linear equations are traditionally divided into (i) direct and (ii) indirect, or iterative,
methods. The direct methods include Gaussian elimination, the indirect
methods include the Gauss-Seidel. The direct methods have the advantages (a) that they will always produce the solution provided that it
exists, is unique and that sufficient accuracy is retained at every
stage, and (b) that the solution is found after a known number of operations. They have the disadvantage that very sparse systems of equations,
such as arise in finite difference approximations to differential
equations, may become rapidly less sparse as the elimination process
proceeds so raising the storage requirement from a multiple of n (for
n equations) to something like n^2 . The iterative methods, on the
other hand, may fail to converge to a solution and if they do converge
it is not obvious how many operations they will require to produce the
desired accuracy. They have, however, the very considerable advantages
that they are very well suited to computers and preserve the sparsity
of the coefficient matrix throughout.

Direct methods of solution of non-linear systems are rarely available;
there is, after all, no direct method for solving the general polynomial
of even the fifth degree and so iterative methods are generally used.
As in the case of linear systems, convergence may not always occur,
though conditions sufficient to ensure convergence are usually known; and
although in some cases the number of iterations necessary to produce
convergence to a specified accuracy may not be easily predicted, it is
frequently not a matter of great importance and, if time is limited,
accelerating techniques can often be used.

The revival of interest in iterative methods brought about by the use
of computers has led to significant advances in the study of functions
which are iteratively defined, e.g. by a relation of the type

$$Z_{n+1} = F(Z_n)$$

where Z_0 is a given complex number and the function $F(Z)$ may
contain one or more parameters. Some functions of this type, such as

$$Z_{n+1} = Z_n^2 + C ,$$

were studied over 60 years ago by Julia (1918) and Fatou (1919), but
attracted relatively little interest at that time. In the case where
the function F(Z) involves one complex parameter C and we define
the set of points K_C to be those points Z such that the iterated
sequence of points given by

$$Z, \ F(Z), \ F(F(Z)), \ \ldots \ etc.,$$

does <u>not</u> go to ∞ , then the boundary of K_C is called the <u>Julia set</u>
associated with F(Z) and C . Only recently, thanks to the avail-
ability of computers and, particularly, of colour graphics terminals
has the extraordinary nature of these Julia sets and their numerous
spin-offs been appreciated.   For example, the Mandelbrot set is
defined as the set of values of C for which K_C is connected. The
boundary of the Mandelbrot set when Z_n is generated by the quadratic
relation above is a <u>fractal</u> curve, the discovery of which, due to
Mandelbrot, has inspired a great deal of exciting and attractive
research by Douady, Hubbard and many others (West, <u>Supporting Papers</u>).

1.6 Algorithms

An algorithm is simply a procedure for solving a specific
problem or class of problems. The notion of an algorithm has been
around for over 2000 years (e.g. the Euclidean Algorithm for finding
the highest common factor of two integers), but it has attracted much
greater interest in recent years following the introduction of
computers and their application not only in mathematics but also to
problems   arising in technology, automation, business, commerce,
economics, the social sciences, etc. Computer algorithms have been
developed for many commonly occurring types of problem.   In some cases
several algorithms have been produced to solve the same problem, e.g.
to sort a file of names into alphabetical order or to invert a matrix,
and in such cases people who wish to use an algorithm will not only
want to be sure that the algorithm will do what it is supposed to do,
but also which of the several algorithms available is, in some sense,
the "best" for their purposes. An algorithm which economises on
processor time may be extravagant in its use of storage space or vice-
versa and the need to find algorithms which are optimal, or at least
efficient, with respect to one or more parameters has led to the
development of complexity theory.   Thus the Fast Fourier Transform has
reduced the time complexity from order n^2 to order n log n , which
is of considerable practical importance for large values of n . More
recently the problem of designing algorithms which can be efficiently
run on several processors working in parallel has attracted considerable interest. Algorithms which are ideal on a single processor
may be highly inefficient, or even fail entirely, on parallel
processors and the design of suitable parallel algorithms for even the
commonest problems is a matter for research.

1.7 Symbolic Manipulation Systems

The possibility of using a computer to manipulate symbols,
rather than numbers, and so provide users with packages for algebraic

manipulation and indefinite integration was appreciated from the earliest days of computers. Packages such as ALPAK and Slagle's SAINT (Slagle, 1963) both date from the early 1960's. Not only were such packages available, they were used. Around 1960, Lajos Tokacs used ALPAK to carry out some very tedious algebraic manipulation involving 1200 terms to find the second moment in a problem in queueing theory, of importance to Bell Laboratories. No-one had had the courage or energy to do this by hand. When the second moment was finally found it reduced to just three terms, after which a shortened mathematical derivation was obtained and a general theory developed. Two points are worth noting: after the brute-force use of ALPAK the nature of the solution inspired mathematicians to find a more elegant derivation - in support of our remark in Section 1.3; secondly, without the use of a symbolic manipulation package it is unlikely that this work would have been done at all.

Another early system, FORMAC, was utilized to help with the solution of the restricted case of the 3-Bodies Problem and, more recently, G.E. Andrews (1979) has used it to check that two 752-term polynomials, occurring in the theory of plane partitions, are identical.

Some symbolic manipulation packages are general, but many more are applications specific. We have mentioned CAYLEY which is widely used for the study of finite groups both at research level and as a teaching aid. Other specific systems include MATRIX, REDUCE (Fitch, 1985), MACSYMA (Pavelle and Wang, 1985); many more traditional algebra systems are surveyed in Pavelle et al (1981). A more general system connecting aspects of logic, mathematics and computer science is Automath (de Bruijn, below).

Whilst many of these systems perform tedious tasks they are not necessarily based on trivial mathematics. Advances made in symbolic integration have been particularly striking. The theory of integration in closed form, originally due to Liouville (1833), was taught in France for about 50 years (to 1880) and then disappeared from Hermite's Cours d'Analyse under pressure from newer material. The arrival of computers re-opened interest in the subject and recent software (Davenport, 1982) solves problems about which G.H. Hardy (1916) said "there is reason to believe that no such (solution) method can be given". The techniques used in integration packages of this type are highly sophisticated and beyond the experience of most users.

Another noteworthy example of a very practical type is a package in which the computer generates a finite difference scheme for the solution of a differential equation within a region, works out the equation for the mesh elements and analyses the Fourier stability of the approximations (Wirth, 1981). This involves both advanced mathematics and advanced programming.

The availability of packages such as those described will not only relieve mathematicians of a great deal of drudgery and encourage them

to attack problems which hitherto looked too intractable but may even
lead to notable new advances as the use of ALPAK and the refutation of
Hardy's remark illustrate.

1.8 Computers and Mathematical Communication
Whilst it affords great personal satisfaction to prove (or
disprove, or conjecture) a result, the mathematical community only
gains if that result is communicated to others. This communication may
take various forms (though the distinctions are not rigid).

epistolary   - where A writes a letter to B communicating the result;

proscriptive - where A writes the result on a wall (literal or
               metaphorical) for others to read;

privately published - the usual form is a departmental technical
               report, whose existence is announced;

publicly published - journals or books.

This communication may be received either directly by the person who is
going to use the result, or indirectly.

The advent of computed-aided typesetting and camera-ready copy has
obviously changed the visual form of mathematical communication
(particularly the publicly published) and its economics. This has consequences for mathematicians (especially editors) who may need to read
the input to such type-setting systems. But computer technology is
capable of changing and is changing, far more than this.

Epistolary. The telephone has not made a major contribution to mathematical communication (though it makes the administration of mathematics and mathematical communication far easier) since it is a very
poor medium for transmitting formulae or diagrams. The telex is rarely
used.

Hence one is forced to the traditional letter. For factors outside the
control of mathematicians, this service has been getting worse over the
years. Mathematicians used to exchange three letters a day between
Cambridge and London, whereas now three letters a week would be more
likely. Not only does it take longer to develop a joint idea, but the
momentum is often lost.

The computer network offers a solution, by allowing communication via
"electronic mail" instead of physical mail. A very high-bandwidth network like the ARPA net is "virtually instantaneous", but lower bandwidth
ones like the CS net that links most computer science departments in
the US, or the JA net that links many academic institutions in the UK,
can certainly provide overnight delivery, and often in a few hours.

As an example of this, Davenport wrote a paper with a text-processing
system in Cambridge (England), sent the result to Coppersmith in New
York, using three networks to do so, and had the corrections and

amplifications a day later.

Proscriptive. In addition to the physical notice boards in one's own department or elsewhere on which one can place proofs (or, more likely, announcements of technical reports containing proofs), computer networks distribute electronic "bulletin boards" to various sites which "subscribe" to them. In some areas of computer science in North America, most results are announced on such bulletin boards.

Private Publishing. This is closely related to the above. Such networks also distribute electronic "newsletters" to individual subscribers, which often contain lengthy articles in draft form, or state conjectures or problems.

Public Publishing. This is the area whose form has been directly least affected. Though there is talk of it, no serious refereed journals distributed by electronic means exist.

All of these methods distribute information to the recipients. Sometimes this will be information that the recipient can use directly. More often, though, the recipient will only want it later, either in the form "I'm sure I saw something on ...." or "Is there anything on ....". The first is hard to answer unless you remember where it was seen. Searching back issues of a particular journal is relatively easy. Finding something that was pinned to a notice board a year ago is almost impossible. If proper archives are kept, finding an item on a year-old electronic bulletin board merely requires programming an editor suitably.

Searching for information on particular subjects is very hard. Classifications such as the AMS one are inevitably too broad, and cause problems on the boundaries, which is where one is most likely to want to look. This is the area of "information retrieval". Already the last eleven years of Mathematical Reviews are on-line, and it is possible to find many of the necessary papers by looking for suitable words among the titles, keywords or reviews. Davenport has used this twice with excellent results when a problem he was working on turned out to be reducible to one in a different field. When this is coupled with citation indices, which are certainly computer produced even if they are distributed on paper, we can find generalisations (or refutations!) of the result as well. An information retrieval system of this type for literature in physics was set up by Kessler at MIT in 1965 (Kessler, 1965) and, for computer science by Churchhouse at the Atlas Laboratory in the UK in 1966 (Churchhouse, 1969). If a database of titles, keywords, references and most importantly of all, automatically generated citation indexes of papers in mathematical journals were available on-line via a network, mathematical research would be greatly aided and time wasted in re-discovery reduced. Any steps to establish and maintain such information retrieval systems should be encouraged.

1.9 The intellectual, economic and social dangers

In one of the working papers for the Strasbourg Symposium, Atiyah drew attention to certain dangers to mathematics which might be associated with the widespread introduction and use of computers by students in schools and universities; this paper is reprinted in this volume. We believe that it is right to be aware of these dangers; but only time will tell how real they are and the situation, in any case, differs from one country to another. We believe also, however, that the benefits accruing from the application of computers to mathematics will far outweigh the dangers, particularly when we are forewarned of what those dangers are.

PART II

THE IMPACT OF COMPUTERS AND COMPUTER SCIENCE ON THE

MATHEMATICS CURRICULUM

2.1 The Common Mathematical Needs of Students in Mathematics,
     Science and Engineering

(a) Preparation for University Mathematics

To provide a context in which to discuss the impact of computers and
computer science on curriculum and methodology, it is necessary to
agree first, in general, on the appropriate mathematics for the secondary school student and then to consider the university curriculum.
Since there are significant differences between different parts of the
world on when secondary school ends and university instruction begins,
the comments which follow will have to be interpreted in the local
context.

Algebra has traditionally been an important subject in high school.
Since elements of abstract algebra are likely to become increasingly
important in mathematics education, it is clear that algebra will
remain of central importance in the secondary school curriculum. The
important thing, however, is not to have students achieve great manipulative skill in algebra (e.g. in polynomial algebra) but rather to
teach them to consider algebra as a natural tool for solving problems
in many situations. Nevertheless, the ability to use formulas and
other algebraic expressions will remain necessary.

In recent years there has been a trend toward replacing much of
Euclidean plane geometry with those aspects of geometry more closely
akin to algebra. This is useful as a preparation for university
mathematics but there is much feeling among mathematics educators that
the loss of Euclidean geometry is a sad development. A consensus on
how geometry might best be taught at school and university is not yet
available. It should be noted, however, that some computer scientists
feel that the aspect of traditional instruction in geometry concerned
with teaching the meaning and construction of rigorous proofs can be
achieved through material concerned with the verification of algorithms.

For many parts of mathematics trigonometry is useful preparation. But
we note that much of the tedious work which was necessary in the past,
both numerical and symbolic, can now be done or will soon be able to be
done on hand-held computers.

Next we mention calculus. In many countries this has been a secondary
school subject for many years for most university-bound students while
in other countries only the very best students begin calculus in
secondary school. The main thrust of secondary school calculus has
been to provide students with techniques, and to prepare those intending to study mathematics at university with a first introduction to the
concepts they will encounter at the university level. Probably much of
the recent work on using computers to teach calculus (see Section 2.2
(b)) is more applicable at this level than at the university level.

Various new subjects have become part of the secondary school curriculum in recent years. Among these is probability which has come into
the curriculum in many countries. From the point of view of this
conference, discrete probability spaces, the binomial distribution and
related topics are more useful than going into statistics which may be
too difficult to teach at this level. (However, an introduction to
data analysis (see Appendix B) is quite appropriate at the secondary
school level.) Another subject, about which there will be further
discussion below, which we would like to see more of in the secondary
school curriculum, is discrete mathematics including elementary counting, number systems other than decimal, the binomial theorem, induction
and recursion. In this connection it would be appropriate to introduce
both the design and verification of a number of important algorithms
such as those for sorting.

What to do about the computer itself for this age group is a difficult
question. Its possibilities for calculus have already been mentioned.
Consideration also needs to be given to its use for any instruction
related to algorithms. Care must be given to avoid emphasis on the
computer as a toy but rather to present it in the context of computer
science as a science.

We might go on to discuss doing almost all the above in terms of models
and practical problems. But we note that the problems of teacher
training for everything we have mentioned are already formidable. We
must learn from the 'new math' experiences of the 1960s and avoid
trying to achieve too much at once.

(b)   The University Mathematics Curriculum

The core of the university mathematics curriculum for many years has
been the calculus and, to a lesser extent, linear algebra. This is the
case no matter how much mathematics the student may have studied in
secondary school. The effect of computers on this curriculum is mainly
one of methodology, not content. That is, the use of computers may
allow more interesting and effective presentations of classical subject
matter but, in and of themselves, computers have little effect on what
subject matter is important to the beginning university student. (An
exception to this may be symbolic mathematical systems ("computer
algebra" systems) whose manipulative power suggests a deemphasis on the
more skill-oriented portions of the curriculum (see Appendix A).)

Informatics (i.e. computer science), however, does imply changes in the content of the core curriculum. This is essentially because informatics is a highly mathematical discipline but one whose problems require almost universally the tools of discrete rather than continuous mathematics. Thus, there is now a strong argument to provide a balance in the core curriculum between the traditional continuous mathematics topics and topics in discrete mathematics (Ralston (1981), Ralston and Young (1983)). For university courses aimed at a broad spectrum of mathematics, science and engineering students, this balance may well contain nearly equal portions of the continuous and the discrete. For those courses aimed at specific student populations, the balance might be weighted more in the direction of the discrete for informatics and social science students, might be about equal for mathematics students themselves and surely should be weighted more toward traditional continuous mathematics for physical science and engineering students. It needs to be emphasized, however, that all groups of students need some exposure to both the continuous and discrete approaches to mathematics. Whether students are exposed to calculus first and then discrete mathematics or vice versa will depend on the student population and on institutional convenience.

The actual content of the discrete mathematics component will be quite variable for some years until considerable experience is obtained with what to teach and how to teach it. We note only that the discrete component will normally contain at least some "traditional" discrete mathematics (e.g. combinatorics, graph theory, discrete probability, difference equations) as well as some abstract algebra although the latter may follow in a later course after completion of the core courses.

We note also the importance of an early introduction to mathematical logic in the core university curriculum. Although traditionally an advanced undergraduate or a post-graduate subject (at which levels there will be a continuing need for specialized courses), logic is so important in informatics that it needs to be introduced early in the university mathematics curriculum (Davenport, Supporting Papers) and even, perhaps, in the secondary school curriculum (see Section 2.2(c)).

As a final matter, we stress the importance of using the paradigms of informatics (e.g. an algorithmic approach, iteration, recursion) in the teaching of mathematics at all levels. Although these paradigms may seem most easily applicable to discrete mathematics, there is considerable scope for their introduction into the classical continuous curriculum.

The reader may be surprised to find no mention of numerical analysis here (or hereafter in this document) because this subject is the one that most obviously combines the continuous and discrete approaches to mathematics. But we take the position that numerical analysis is now such a well-established subject in the mathematics curriculum that it does not need to be discussed in the context of this report. This is,

however, not to say that the subject matter of numerical analysis is no
longer affected by advances in computing; developments in, for example,
parallel computing will have great impact on numerical analysis.

We now proceed to give considerably more detail than in this introduction about the specific curriculum areas discussed above.

2.2 A Discussion of Particular Curriculum Areas on which Computers and Informatics have an Impact

(a) Discrete Mathematics Courses

We begin with a consideration of what topics in discrete mathematics are
essential to the beginning informatics student and, as well, are important to the mathematical development of a variety of other students
including mathematics students themselves. It should be noted that in
many countries some, if not most of these topics are already part of the
precollege curriculum. Where this is not the case, as in North America,
there is a growing movement to introduce these topics into a single
course, most often in one semester but sometimes in a full year course,
with the title Discrete Mathematics (Ralston and Young (1983), MAA
(1984)). Although the topics to be listed below cover a broad spectrum,
it is possible to design a coherent course covering these topics if the
course is based on the theme of trying to understand the applications of
these topics to computing. This means an emphasis on algorithms and
their analysis as well as on such other fundamental concepts as mathematical induction and the representation of mathematical constructs by
functions and relations.

In the syllabus for a Discrete Mathematics course which follows, we have
taken into account the needs of computer science students but have also
used our perceptions of what mathematics faculty in most countries are
ready and willing to teach (see the papers by Bogart, Dubinsky, and
Seda, Supporting Papers).

A Discrete Mathematics Syllabus

1. Sets and set operations, relations, equivalence relations, partial
    orderings, functions.

2. Elementary symbolic logic including the standard logical connec-
    tives, conditional expressions and an analysis of their meaning;
    the concept of proof, at least as a convincing explanation; the
    relation to the contrapositive form of a conditional statement;
    proof by contradiction.

3. The principle of mathematical induction and its application to
    recursive definitions.

4. Basic counting techniques including the sum and product rules;
    binomial coefficients; permutations, subsets and multisets

(i.e. combinations with and without repetitions); the principle
of inclusion and exclusion.

5. Difference equations (i.e. recurrence relations); equations with
    constant coefficients; first order equations; the relationship
    of recurrence relations to the analysis of algorithms.

6. Graphs, digraphs and trees; path and connectivity problems; tree
    traversal, game trees and spanning trees; trees as fundamental
    data structures.

7. Discrete probability including random variables, discrete
    distributions and expected value.

In addition, other possible topics depending upon local needs and
desires are

8. Matrix algebra including matrix operations, inverses, deter-
    minants; linear programming; applications.

9. Number systems, particularly the discrete systems used on
    computers.

10. Algebraic structures such as rings, groups, etc.

11. Finite state machines and their relation to languages and
    algorithms.

And, of course, there can be extensions of all the above topic areas to
more advanced subject matter if desired and appropriate.

The experience of those who have taught such discrete mathematics
courses is that, despite the potpourri of topics listed above, these
courses can be made interesting and satisfying if a consistent,
coherent approach is taken as suggested above (see papers by Bogart,
and Jenkyns and Muller, Supporting Papers).

Following a course from a syllabus like that above, a variety of
advanced courses in discrete mathematics can be contemplated although
only the largest institutions would be able to offer all of these.
Indeed, each of the 11 subject areas listed above suggests one or more
advanced courses which would build on the introductory material in a
first discrete mathematics course. Most of these courses are currently
in a process of evolution as the subject matter in the first discrete
mathematics course changes and develops and as the applications of
discrete mathematics grow and diversify. A program which combines a
carefully constructed introductory discrete mathematics course with
several advanced courses will give the student a firm basis for studying informatics as well as providing a basis for professional work in
modern applied mathematics and other fields in science and engineering.

(b) Calculus in the Computer Age

I. The Role and Relevance of Calculus

Among the key factors which compel change in the teaching of university
mathematics courses are:

- the substantial experience with minicomputers and microcomputers
   and programming packages which many students have had before
   coming to the university;

- the growth of new areas of applied mathematics such as the
   analysis of algorithms and computational complexity.

One result of this is that many students have attitudes and expectations
which lead them to believe that the most challenging and meaningful
mathematical problems today are related to computers and informatics.
This cannot help but influence how we must motivate mathematics
students and all other students in mathematics courses.

In considering the place of calculus in the computer age, we cannot
forget that it is one of humankind's greatest intellectual achievements,
one of which every educated person should be aware. Its history exem-
plifies the "unreasonable effectiveness" of mathematics better than any
other branch of mathematics. And its effectiveness is as great today
as it has ever been. But this does not excuse teaching calculus as is
so often the case now with an emphasis only on the execution of mechanical procedures - and paper-and-pencil procedures at that. Instead
calculus needs to be taught to illustrate the unique ways of thinking
it epitomizes

The realm of applications of calculus remains immense. Applications of
calculus may even be increasing due to the increasing mathematization
of heretofore qualitative sciences like biology. In constructing
calculus models of phenomena and then solving the resulting equations,
there is often an interplay between these models and their discrete
counterparts with the calculus models representing the limiting behaviour of the discrete models. It is now more important than ever to
include this interplay in calculus (and discrete mathematics) courses
because inevitably the solution of most problems in calculus involves
the (discrete) computer (Winkelmann, 1984a). The discretization necessary to solve problems of calculus with a computer often has not borne
a close relationship to the underlying discrete model. But the
increasing power of computers means that more and more frequently it is
possible to have computer models which mirror very closely the discrete
models from which the continuous model was initially abstracted
(Winkelmann, 1984b, Supporting Papers).

There already are powerful software tools which can be used in the
study of calculus. These include symbolic mathematical systems (see
Appendix A) and a variety of graphical packages. Advances are taking

place so rapidly in these areas, however, that it will soon be the case
that very powerful symbolic and graphical systems will be available on
microcomputers and even on hand-held computers. One result of this is
that an understanding of functions, variables, parameters, derivatives
etc. and the ability to interpret formulas and graphics is becoming
more important to the student than skills in executing the (numerical
or symbolic) procedures of calculus. In the teaching of calculus to
all students the need is clear for a shift from an emphasis on calcula-
tional technique to one which emphasizes the development of mathematical
insight (see, for example, Murakami and Hata, below).

II.   The Content of Calculus Courses

If functional behaviour and representation are to be the focus of the
calculus course, then continuous functions and discrete functions (i.e.
sequences) must be emphasized and motivated by a wide variety of
mathematical models. Sequences should be defined iteratively and
recursively. (Note: Some would argue that sequences belong more
properly in the discrete mathematics course discussed in the previous
section. This only illustrates the need to bring the discrete and
continuous points of view together into an integrated sequence of
courses as soon as possible (see, for example, Rice and Seidman,
below).)

An important theme in calculus courses should be the contrast between
the local and global behaviour of functions. Local behaviour is, of
course, derived by studying the derivative for continuous functions
(and the difference operator for discrete functions). And similarly
the integral (and summation) operators are used to derive global
information about functions. Undoubtedly it will remain necessary to
develop some ability to do formal computations with derivatives and
integrals. But the major emphasis should be on numerical algorithms
(particularly for integrals) and on how derivatives and integrals can
be used to understand the behaviour of functions.

A topic such as the Taylor series representation of a function should
be used to show how good local information can be obtained using low-
degree polynomials. A valuable comparison in this context would be
between Taylor polynomials and interpolating polynomials, another area
where the analogy between the continuous and the discrete may be usefully shown. The use of the mean-value theorem as a tool to estimate
errors suggests another way in which a computational approach changes
the perspective on classical topics in calculus.

Finally, there should be a balance in the calculus course between
traditional topics and ones whose importance has greatly increased
because of the advent of computers and informatics. Thus, for example,
the O( ) and o( ) notations are not always taught in calculus
courses, but they should become so.

This discussion is intended only to provide the flavour of how an orientation toward computation should change the approach toward teaching most of the standard calculus topics.

III. Computers for Learning and Teaching Calculus

Computers enable teachers to modify their methods of teaching calculus (and, of course, much other mathematics also) in order to meet better the need of their students. Computer graphics is a powerful medium in which to provide examples - and non-examples - of continuous functions, discontinuous functions, the area under a curve, direction fields and nowhere differentiable functions as well as in many other areas. Well-designed software (there isn't nearly enough of this yet) can be used by students to discover and explore the concepts mentioned above as well as such fundamental concepts as slope and tangency (see also Section 2.3). But the effective use of such software requires that teachers sometimes depart from a lecturing style and go instead to a guiding and interacting style.

Well-designed software will also permit enhancements by students through the writing of (usually short) programs. This is just another way in which students can be actively involved in their own learning although it is important that the use of the computer does not become the message instead of the mathematics which it is supposed to illustrate (see Tall and West, below, Tall (1985)).

Another impact of the computer in calculus may be to change the order in which topics are taught. For example, it may not be best to introduce limits at the very start of a calculus course. Tangent functions and area under a curve can be motivated and defined graphically. When a formal definition of a limit is needed, students will be ready for it. As another example, differential equations can now be treated much earlier in the curriculum than was previously possible because of the ease of understanding made possible by new graphics systems. They could be introduced right after differentiation and before integration. We need to study whether such reorderings will lead to a greater or more rapid understanding of fundamental concepts and theorems.

To take full advantage of the use of computers in teaching calculus, it will be necessary to change the standard classroom environment. Classrooms need to be provided with large monitors or screens on which the monitors may be projected. Outside the classroom, students need administratively easy and user-friendly access to computers and software. Teachers will need private computer facilities to be used to prepare course material. A prerequisite for this is in-service training so that teachers may become comfortable with computers and then fluent in their use and aware of possibilities beyond what may be available in the particular software on which they have learned (Winter, Supporting Papers).

(c) Logic for Mathematicians and Computer Scientists

It has been argued above that mathematical logic needs to be taught at
a lower level than has been customary because of its importance in the
education of computer scientists. Because this will be a new idea for
many readers, some details on what is desirable in this area are
contained in this section.

If a logic course requiring considerable mathematical maturity is not
the right thing, then neither is the kind of course which explains
logical inferences by means of examples in a natural language. What
we need instead is a course to teach beginning university students the
essential rules of the game of mathematics. Two reasons why we need
such a course are that:

- in the computer age large numbers of people, particularly in
  scientific and technical professions, will need to handle
  statements of a logical or mathematical nature in a very
  precise fashion.

- now that the teaching of proof in geometry is rapidly
  disappearing in many countries, we need some place in the
  curriculum where students learn the art of proof.

The tools of the working mathematician can be explained through the
use of the natural deduction style with the propositional and predicate
calculus (Fitch, (1952)). A course for first-year university students
of informatics at the University of Eindhoven takes this approach
(de Bruijn, Supporting Papers). Students learn how to arrange a proof
and how to deal with naive set theory, predicates, bound variables and
quantifiers. They were able to understand the mechanism of indirect
proof, proofs by induction and notions like uniform convergence.

A good way to start with the art of proof is through the derivation of
formulas in the propositional calculus using only the implication and
conjunction connectives with the usual elimination and introduction
rules. Even at this stage students can be given exercises which
require some ingenuity.

At this level it is important to present the material as a bag of tools
and not to try to prove statements in a metatheory since this would
surely confuse beginners. Even so, there should be some introduction
to truth tables in such a course to show students how some results
which were obtained with natural deduction can be obtained in another
way.

A logic course for beginners should begin with a discussion on syntax
in which the students learn to represent formulas (including those with
bound variables) as trees and in which they learn to handle substitution. This portion of the course should also include instruction on
how to handle parentheses, precedence rules for operators and prefix

and infix notation. This material could precede introduction to a
computer language but could better serve as a formal counterpart to
topics in a programming language course.

A course such as the one described here might well be prescribed for
mathematics as well as informatics students.

2.3 Exploration and Discovery in Mathematics
The idea of using computers to enable students to explore
mathematics and discover mathematics for themselves has been mentioned
already. However, the advent of powerful and available computer systems
makes this point so important in teaching mathematics today that we
devote an entire section to it (see also Murakami and Hata, below).

First, why should exploration and discovery be important components of
the educational process in mathematics? The answers parallel the
reasons why we teach mathematics in the first place:

-   active learning leads to better retention and understanding and
    more liking of the mathematics we teach because the mathematics
    is seen as a basic component of human culture; it also leads to
    more self-confidence in the ability to use mathematics to solve
    problems;

-   exploration and discovery helps to teach people to think;

-   discovery provides the greatest aesthetic experience in
    mathematics, the "aha" of seeing or proving something, is
    what makes mathematics attractive;

-   exploration and discovery are perhaps the best ways for
    students to see that mathematics is so useful;

-   discovery enables the student to see a familiar idea applic-
    able in a new context, thereby enabling a grasp of the power
    and universality of mathematics.

Computer technology may be used to assist in mathematical exploration
and discovery in a variety of ways; for example:

-   through visualization of a great variety of two and three
    dimensional objects via computer graphics, students may
    explore questions and discover results by themselves (Tall,
    1985);

-   through computer graphical presentations of interesting
    geometries like "flatland" and turtle geometry;

-   via exploratory data analysis (see Appendix B below);

- by graphical and numerical explorations of how to approximate complicated functions by simple ones;

- by applying the first step of the inductive paradigm - compute, conjecture, prove - in many, many different situations;

- by using symbolic mathematical systems (see Appendix A below) to discover mathematical formulas such as the binomial theorem;

- by designing and executing different algorithms for the same or related tasks.

This list could be made much longer. Readers will probably be led to make their own suggestions.

There are various implications to using computers to facilitate exploration and discovery:

- we must start with easy tasks so that students feel they are really succeeding on their own and are not being led step by step by the teacher;

- teachers need to be trained for this kind of instructional mode; few teachers can handle these ideas without training; and, in particular, testing what has been learned by the student is not easy.

But experience has shown that success is not only possible but yields rich rewards. The difficulties can be overcome; teachers can be trained to feel comfortable with this mode of learning.

PART III

COMPUTERS AS AN AID FOR TEACHING

AND LEARNING MATHEMATICS

Introduction

Mathematicians and mathematics teachers have been provided with a new tool, the computer. There is no shortage of applications or interesting examples which one can quote. But, like all tools, the computer by itself does not supply the solution to our problems, not least the problems of mathematics education. There is no automatic beneficial effect linked to a computer: the mere provision of micros in a class- or lectureroom will not solve teaching problems.

It is essential, therefore, that we should develop a serious programme of research, experimentation and reflective criticism into the use of informatics and the computer as an aid. It will not suffice to think only in terms of mathematics and the computer, and of the production of software which amuses and interests mathematicians. We must also take into account types of knowledge and the ways in which these can be transmitted, and attempt to study, in a serious epistemologically-based manner, various concepts and the obstacles which they present to learners. We must think of students, their development and the matching of new and old knowledge. We must consider in depth those teaching possibilities created by the computer. It is essential, above all, that we should move beyond the stage of opinions, enthusiasms, and wishful thinking and engage in a true analysis of the issues. Only in this way will we come to a true resolution of certain problems of teaching. Such research, of necessity experimental, will have to be critically evaluated. It must be shown how, in given circumstances, the use of the computer can facilitate the acquisition of a particular concept. Finally, such research work will have to be built upon and developed to provide a vital component in the training (whether formal or self-directed) of teachers and lecturers. Only then can computers have any large-scale effect on mathematics teaching.

3.1 A changing view of mathematics

There are many references in this book to the way in which the computer can lead to a changed view of what mathematics and mathematical activities comprise. For example, the experimental aspects of mathematics assume greater prominence (see Section 2.3), and there is a corresponding wish to ensure that provision should be made for students to acquire skills in, and experience of, observing, exploring,

forming insights and intuitions, making predictions, testing hypotheses,
conducting trials, controlling variables, simulating, etc. Examples of
how such work can be carried out are to be found later in this book and
in the volume of Supporting Papers. Much, however, remains to be done.
Thus, for example, a mechanism needs to be found for disseminating
information about fruitful experimental environments and how these can
be formed.

Yet, when we put new emphasis on those particular activities listed
above, it is, nevertheless, necessary to ensure that such traditional
activities as proving, generalising and abstracting are not neglected
or omitted. We will need to find an appropriate balance between
'experimental' and more formal mathematics.

The possibilities presented by the computer (see, for example, those
described in the paper below by Lane et al) will actually help focus
our attention on the kind and types of knowledge which we wish students
to acquire. Not only new possibilities are offered to us, but also a
greater incentive more precisely to identify our educational goals.

If our aims of teaching change significantly so as to encompass and
stress more the 'process' of mathematics, rather than the 'products' of
the mathematical activities of others, then there will, of course, be a
need to identify those parts of mathematics most suitable for our
purposes. Topics and areas of mathematics must be selected which
encourage and facilitate an experimental approach.

Finally, in this section we must stress two important, inter-related
points. Many, indeed the majority, of our students may not intend to
become mathematicians. We must not lose sight of the implications of
this in terms of educational goals and emphases. Many of these may be
students of the experimental sciences. This raises further important
issues, for experiments in mathematics differ somewhat from those in
the physical and natural sciences. The techniques are often very
similar, but in mathematics we have that extra, vital ingredient of
'proof'. Experiments are an essential and neglected part of mathematics, yet mathematics is not an experimental science. The distinctions between disciplines and ways of thought will have to be displayed
and observed.

3.2 Computers change the relation between teacher and student

Computers can affect the behaviour of students. This
creates new interactions and relationships between student, knowledge,
computer and teacher. The role of the teacher in such situations
demands considerable thought.

(a) The mathematical activity of the student

Students will be better able to learn conceptual material and develop
autonomous (as opposed to imitative) behaviour patterns with respect to
mathematical ideas, if they can be cognitively active in response to

mathematical phenomena presented to them. This activity should consist
of the formation of mental images to represent mathematical objects and
processes. It should also include the development of skills in manipulating them. In this way students can increase their ability to think
mathematically.

Inducing students to emerge from passivity and to think actively about
mathematics is, however, not easy. One approach is to make use of the
computer to supply sufficiently powerful and novel experiences to stimulate such behaviour. The action of a computer program and the structure
of data as it is represented in the computer can form useful models for
thinking about mathematical entities. For example, a "WHILE loop"
whose body is a simple sum is a process that can represent the mathematical entity

$$\sum_{i=1}^{m} x_i \; .$$

This expression, which troubles so many students, can then be thought
of in terms of a simple, familiar and useful computer process. Again,
in PASCAL, representing a fraction as a record with two integer fields
(the second being non-zero) helps students think about rational numbers
as ordered pairs of integers, especially if they are given the experience of writing programs to implement the arithmetic of fractions
without truncation.

Many examples of ways in which such experiences can be incorporated
into mainstream, tertiary-level courses have already been given (see,
for example, Supporting Papers). Moreover, the success of such initiatives would seem to be independent of several issues which in discussion tend to be over-rated. An important factor in this approach
appears to be that students should write the programs and so must be
cognitively active about the processes and data structures they are
implementing. These experiences are then coordinated with classroom
activity.

Dubinsky (Supporting Papers) describes a course which involves such
discrete mathematics topics as quantification of predicates and induction. The programs are written by students in a very high level
language (SETL) running on a mainframe. The printed results of the
program form the basis for reflection and discussion.

Mascarello and Winkelmann (below) describe a course containing
'continuous' topics such as multiple integration and ordinary differential equations. Here the students wrote programs in a low level
language (BASIC) running on a microcomputer. These were interactive
and the results were used for experimentation and demonstration.

Of course, writing programs is not the only useful way in which students
can use the computer. The use of complicated software packages for
illustration of phenomena that are very difficult to display otherwise

can clearly broaden the students' awareness and add to their general understanding (see, for example, the paper below by Tall and West). They can, of course, also be used for exploration and discovery (see Section 2.3 above). Indeed some would see the most exciting opportunity offered by the computer to be the way in which it can motivate students to exercise the process of discovery. Here we should only stress the need to see exploration and discovery as essential mathematical activities to be practised. Traditionally, this has not been so - teaching and learning have been almost wholly concerned with the transmission and reception of accepted mathematical facts. However, now, for example, computer symbolic maths systems permit such rapid and flawless processing of non-trivial examples that it is easy first to look for patterns which suggest conjectures and generalisations, and then to search for counter-examples or machine-aided proofs (see, for example, the paper by Lane et al).

Computers then can greatly assist us in extending the range and the depth of students' mathematical activities. In some approaches the students will write their own programs (and there will be an attendant risk that mathematical aims may become obscured by some of the programming problems); in others students will use prepared software. Both approaches have already been shown to be of great value: further investigations into both will now have to be undertaken.

(b)   The role of the teacher

The computer can be used in two distinct ways in the classroom. In one it is an aid for the teacher, an electronic blackboard - more powerful than the traditional blackboard, the overhead projector, or a calculating machine - but nevertheless a tool whose output is almost entirely under the teacher's control. In this role the computer does not upset the traditional balance in the classroom. It will still demand effort on the teacher's part to select or provide suitable software and it can give rise to extra administrative problems; in return it should enhance learning. However, it will not revolutionise the classroom.

If, however, students are allowed and expected to interact with computers then the position changes, for this leads of necessity to a change of methodology. The teacher no longer has total control - his/her role can no longer be limited to exposition, task-setting and marking. The format 'lecture-examples, homework-exam' must be augmented by, for example, 'project - interaction between student, machine and teacher - assessment on the basis of a completed (and possibly debugged) assignment'.

Such a change would produce a revolution in most class- and lecture-rooms. It demands that teachers should not only acquire new knowledge, skills and confidence in the use of hardware and software, but that they should also radically change their present aims and emphases, and accept a lessening in the degree of control which they presently exert over what happens in their classroom. This last demand means a sacrifice of traditional security, at a time when teachers will still be fighting hard to gain new skills and acquire confidence in them. It would be foolish to underestimate the challenge this presents.

The acquisition of new skills will be time-consuming and constantly changing hardware and software will make the process a continuing one. For many mathematicians these new skills will be readily usable in their research work. Others may be tempted - particularly when universities and other educational institutions are under pressure - to feel that such time would be more profitably spent in increasing personal research output, rather than in improving their teaching, particularly if this requires such a large step in the dark.

The participants at Strasbourg were far from optimistic. Many contributors to the conference reported that computer usage was actively avoided by their colleagues. The problems at tertiary level were seen as being particularly great, for the gulf between the traditional lecture often given to a hundred or more students and the classroom/laboratory in which students interact with computers is enormous. To bridge this gulf will need considerable investment in both material and human resources. Time, assistance and in-service training will have to be provided on a scale unprecedented at this level. Particular attention will have to be directed at those teachers who still have many years - even decades - to go before they retire from teaching. First, however, the necessity for change will have to be accepted, and this will only come through clear, unequivocal demonstrations of the benefits which can accrue from innovation.

3.3 Some particular uses of the computer in the classroom

We have already remarked on the way in which computers can assist in the introduction, development and reinforcement of mathematical concepts, in building up intuition and insight, etc. In this section we look at particular ways in which they can be used within the classroom.

(a) Graphic possibilities

Many of the applications of computers in teaching make use of the possibilities provided for graphic display. There is no doubting their value in providing in reasonable time good quality graphic illustrations which can help build intuition, for example, Hubbard and West (Supporting Papers) describe convincingly how this has been done with ordinary differential equations such as $x' = x^2 - t$, whose solutions cannot be written down in elementary terms. Moreover this allows them to discuss exciting questions concerning the behaviour of solutions.

Where the computer scores over many other media is that graphics capabilities now enable movement as well as static diagrams to be portrayed. This, of course, was true of the film. Yet now the possibility of being able to change parameters adds a completely new dimension to the teaching/learning experience.

Many examples and illustrations were produced at Strasbourg of visual representations from areas such as calculus, differential equations, linear algebra, and numerical analysis. Some of the examples were for teacher use only whilst others were of a self-teaching or interactive design.

(b)   Self-evaluation and individualised instruction

The computer can provide a tool for self-evaluation and can help the student to take charge of the organisation of his own work. It is a difficult problem for students to judge how well they are coping with a subject. One use of computers is to enable students to test themselves. Question banks can be made available and instantaneous scores given (see, for example, Tait and Hughes (1984) which gives examples drawn from other disciplines).

The advantages of Computer Assisted Learning for individualised instruction have, of course, been argued for some twenty years: that the computer can offer non-threatening, individualised responses to students. There have, indeed, been several demonstrations of the value of CAL, for example, PLATO in the USA. However, as the cognitive complexity of what has to be learned increases then the difficulties of producing adequate software become very great. It was noticeable that no examples of individualised learning programmes for use at a tertiary level were described or exhibited at Strasbourg.

The problems become less pronounced when the aim of the program is to revise and not to teach. Thus 'Recalling Algebra' and 'Recalling Mathematics' [Kinch] are examples of software designed to help students prepare for the Entry Level Mathematics Exam at California State University which have been favourably received.

(c)   Assessment and Recording

The use of the computer for testing students' progress is described in Bitter (Supporting Papers). He employs the random generation of test items. Such testing can, of course, go far beyond reliance on multiple choice items, and can measure responses other than correct and incorrect. Such newer testing procedures, which can be designed to capitalise on the graphic potentialities of the computer, can reduce testing time, allow tests to be broken off and resumed at any time, offer immediate summaries and analyses, and assign specific help on identified deficiencies.

The obvious disadvantages include preparation costs and the need to provide ready access to a computer. Open-ended testing of projects or personal problem solving is at present difficult, but beginnings are being made.

Computer assisted recording also has great potential.

(d)   Student errors

Related to the possibilities described above is that of investigating
the errors which students make in learning mathematics.   Such information can be used in two ways: to help the student remove misconceptions,
i.e. the role in which it is used in individualised CAL, or to help the
mathematics educator to identify specific points of difficulty and to
design curricula with these in mind.   Errors are symptoms which allow
us not only to identify stumbling blocks, but also to form an impression of the student's conceptions.   The computer allows the student to
respond to his errors in a new way: he can identify and control them
himself and getting rid of them can even become a motivation for
learning.

One example of the use of the computer to detect and correct errors is
found in Okon-Rinné's courseware.   This enables a student to choose a
basic function such as $f(x) = |x|$ and then to experiment with the
effects which translations and reflections have on it.   Thus the graph
can be translated vertically or horizontally or reflected in the
vertical axis.   Simultaneously the function changes to correspond to
the new graph.   The intention is to detect such common errors as con-
fusing $f(x) = |x-2|$, with $f(x) = |x+2|$,   or $f(x) = |x+2|$ with
$f(x) = |x|+2$.   When an error is detected a tutorial subroutine is
activated and afterwards the students have the option of continuing or
branching back to an earlier unit.

3.4   Student responses to work with computers
Many comments were passed at Strasbourg concerning the
enthusiasm generated in students by computer-based systems.   It was
claimed that this had resulted in many students developing a new
interest in the subject and that the general level of student activity
had increased as a result of reacting with a computer package.   Not
only had activity increased, but so had confidence.   Dubinsky typically
reported (of a course on discrete structures):
  'this approach makes for a lively course in which students are
responsive in class and active outside class.   In comparison with
similar groups to whom I have tried to teach this material, these
students seem to be more prone to speak in terms of sets and less
confused by complicated logical statements'.

It must not be thought, however, that enthusiasm can be automatically
generated through the use of a computer.   Much will depend on the
students and the teaching situation; there are also negative experiences to report!   One must also judge on how much students learn as well
as the enthusiasm they show whilst engaged on the task.

Here one is faced with a new problem in one's teaching.   Students can
frequently appear fascinated by computer demonstrations or by working
interactively with a computer, but what happens 'when the machine is
switched off'?   Will the students only be able to imitate what they
have seen or will they obtain a deeper understanding of concepts?

It was recognised that the value of much computer work was largely
dependent upon the follow-up activities which 'must guard against the
possibility that the machine is doing all the work and providing all
the answers'. Many traditional activities will still have to be
carried out, thus suggesting yet again that the computer's main contribution will be to enhance student understanding and not to save time
for the lecturer. The introduction of the computer is unlikely to
solve (or even ease) the problem of over-loaded syllabuses.

Yet another aspect of student response comes into play when students
engage in activities with computers. What happens to pupil/pupil
interaction in the classroom? There was no evidence at Strasbourg that
much attention had been paid to this important question. Here is a
field for further research.

3.5 The provision of software

Current software resources for our target groups may be
considered in three categories:

(a) Sophisticated systems (in computer terms) such as the symbolic
manipulation systems, large statistical packages, etc., form the first
category. These systems have been developed in a 'goal-oriented'
fashion, that is they seek to provide solutions to specific mathematical problems. They have not needed to consider to any great extent
'pedagogical design'. Interest in their use as pedagogical tools is
growing (see, for example, Newman, Supporting Papers) but experience
so far is limited.

Commercial companies exist with an interest in marketing this type of
software and research mathematicians are involved in creating such
systems. As a result, sophisticated packages are self-perpetuating –
they will exist – we need to understand their pedagogical uses and the
possibly dramatic effects they could have on current mathematics
education.

(b) Less sophisticated in computer terms but still very demanding in
pedagogical design are the software packages suitable for use on a
microcomputer. These packages attempt to aid the students' mathematical
development and employ such themes as visualisation, simulation,
exploration and problem-solving. They may be used by students working
alone, in groups, or with a teacher. Many individuals and groups are
writing such packages. However, such resources are not self-supporting
in commercial terms and, as a result, are not easy to obtain. Because
of their restricted use in limited contexts, it is still the case that
very little is understood about their effect on teaching and learning.

A major problem arises here. The production of packages that can be
recommended for widespread use as pedagogically sound and well-tested
is an expensive, complicated task requiring considerable professional
resources. It should involve fundamental research based on the structured observation of the materials in use in parallel with the develop-

ment of the materials. Thus the team may need to include mathematicians, educators, psychologists, computer scientists, graphic designers, publishers and editors.

The financial needs of such a group would be considerable.

(c) General purpose programming languages can be used as tools towards students' mathematical development and are a readily available teaching resource. Extension of such languages or even creation of new ones expressly for this purpose would be welcome.

This brief consideration of the present position points out the need:

 (i) to establish channels of communication so that researchers and educators are aware of resources currently available;

(ii) to set up structured research studies using currently available resources in order to gain and share understanding of their use as pedagogical tools.

The emergence of software packages has raised a new problem for mathematics teachers, that of black boxes, for they often/usually produce answers without giving any hint of the way in which they were obtained. As Davenport (below) shows, this may well conceal a wealth of deep mathematics. (It could, of course, be argued that the problem is not new, but merely heightened – for students have been employing algorithms whose workings they did not understand for centuries!)

How can students learn (be taught/encouraged) to look critically at the answers supplied? How much should students be required to know about the workings of black-boxes before being allowed to use them? For example, there are packages which invert matrices. If such a package uses floating-point arithmetic, it can give answers which should not be accepted at face value. At least students should be warned about this or, better, should learn to recognise when this has occurred.

3.6 Cultural, social and economic factors
We have written of the computer as an aid to mathematics teaching and learning. So is the overhead projector. The difference though between the two tools is not, however, solely the enormous difference in the range of possibilities opened up by the former. Equally, it springs from the enormous effect which the computer is having upon society outside the confines of educational systems. As a result society has expectations concerning computers and their use – expectations which often have little basis in reality. Students too have expectations about their use. There are then enormous pressures on educators at all levels to use computers, not necessarily for their intrinsic value, but because society expects it, and not to do so might be considered old-fashioned and reactionary.

It will be difficult for computers to be used effectively in education until society has become better informed about their power and limitations. Unrealistic expectations must be strongly discouraged. There is a danger that false advertising by computer companies and software developers, and a pressure from various sections of society could lead to ill-designed, over-optimistic innovation and, in turn, to a backlash comparable with that of the 1970s resulting from the hasty introduction of 'New Math'. Political decision makers in some countries are 'pushing' computers and computer-related curricula into education without adequate consideration of objectives and consequences.

It is important, therefore, to realise that:

- reasonable use of computers in education requires software programs and packages the educational standards and qualities of which are comparable with those technical ones offered by the available hardware;

- integrating computers into the curriculum must be coordinated with teacher/faculty in-service, professional development programs;

- educational budgets must be prepared to permit appropriate expenditure on hardware, software, and teacher development;

- no curriculum should remain stagnant for a long period.

Not all problems associated with computers in education can be anticipated. Many questions need to be answered through research initiatives directed at investigating the possibilities, limitations and possible dangers of computer use in education. Some causes for concern are:

- uniformity in students' thinking and reasoning could arise from overuse of computers in the learning process,

- standardisation of software development (in an attempt to form a commercial market) may lead to mediocrity and conformity,

- subtleties of communication between teachers and students could be impoverished by over-using computers,

- insensitive working with computers could adversely influence the total intellectual development of students (of their intuitive thinking, creativity, perception, etc.).

There were few representatives at Strasbourg from developing countries. Yet their position demands special attention. For them the provision and maintenance of hardware creates great problems. Moreover, scarce resources must be husbanded carefully. The computer could, offer special advantages to them; on the other hand the absence or shortage of computers could widen still further the gap between them and the developed countries.

3.7 Conclusion
We are only experiencing the very beginning of the effect
of computers on the teaching and learning of mathematics. Gradually,
we are beginning to take advantage of some of their more obvious possibilities such as their quick and accurate production of graphical
material, quick and accurate (though not always precise) arithmetic,
quick and accurate algebraic manipulations, their ability to handle and
analyse large quantities of data (which can be employed in learning
systems of the sort described by Allen et al (Supporting Papers)).

One can see in numerous publications and, in particular, in the papers
contributed to the Strasbourg meeting, many examples of mathematical
situations which the computer and informatics allow us to see and
approach from a new point of view. Obvious examples which spring to
mind are the many applications in statistics (dealing with vast quantities of data), in probability (with all the possibilities opened up for
simulation by pseudo-random generators); in geometry too there is a
range of interesting activities: production of images, curve plotting,
the transformation of images (translations, reflections, ...), loci,
exploration of images and figures. The dynamic aspect dominates here:
one can visualise instantly the effect of varying a parameter. In
linear algebra, an algorithmic approach furnishes a tool both for calculations and also for demonstration. Here again the dynamic aspect
plays an important role: to see a matrix steadily assume a diagonal
form is very different from obtaining the result once and for all after
a long and involved calculation. But it is above all in analysis that
the opportunities to utilise informatics are richest and most numerous.
The study of numbers, of functions, of the solution of equations,
observation and study of sequences and series (and in particular of
their speed of convergence), integral calculus, differential equations,
asymptotic expansions, discretisation, power series for functions, ...
In addition to these 'classical' fields where the use of the computer
arises naturally, one has also seen developments in newer fields which
have occurred largely because of computers: formal symbolic logic is a
striking instance; discrete mathematics can provide us with other
examples. The computer is not only an aid for computation and demonstration, but a force for development.

In all of these cases, the contribution of the computer takes several
forms: firstly it is a calculating tool allowing numerous and rapid
calculations; it also serves to place renewed emphasis on numerical
methods, and thus on the study of algorithms; and, especially, it is a
pedagogical tool, for promoting teaching and learning.

However, let us reiterate, the act of using a computer does not automatically lead to an improvement. It is not a magic wand! Like all
tools, it can serve us badly; we must learn how to get the best from it.

Computers are now widely to be found in schools and universities.
Teachers are being trained in their use, but principally in techniques
and programming. Much remains to be done before we can give a true

pedagogical training. It is also necessary to bear in mind that if we wish to change the educational system, then there will be a need simultaneously to reform both the training given to those preparing to teach in schools and universities and also the continuing education of existing teachers.

At the same time there is the need to carry out much research and experimentation so that we may effectively understand and control the impact of the use of the computer on students' learning and on their conceptions and representations of mathematical objects. Only after such studies will we be able to provide high quality software and, most importantly, a new range of didactical activities, tasks and situations to enhance learning.

APPENDICES

A. Symbolic Mathematical Systems

The best known such system for large computers is Macsyma and the best known for microcomputers is muMath. These systems do <u>symbolically</u> the standard processes of secondary school and college algebra and of calculus. Thus, they differentiate, they integrate (definite and indefinite), they do polynomial algebra, they do infinite precision rational arithmetic, they solve linear systems and quadratic equations – all symbolically although they will provide normal numerical answers, too, when the user wishes. Advanced systems like Macsyma will also do many advanced mathematical processes such as contour integration and tensor calculus calculations. But even muMath is powerful enough to do almost all of the manipulations students need to do through the first year of a calculus sequence (see, for example, Lane and Stoutemyer, below).

The most important point to note for this document is that these systems are rapidly becoming more powerful and will soon be available on handheld computers. Thus, it is becoming increasingly hard to justify trying to teach students to become good symbol manipulators unless it can be shown – but no one yet has so shown – that such skills are required in order to develop an understanding of the underlying mathematics at whatever level such understanding is desired. Of course, developing such understanding is as – or more – necessary in mathematics education than it has ever been.

B. Exploratory Data Analysis

This is a subject which has become pedagogically <u>possible</u> for schools only in recent years. It could not be handled in a classroom until calculators and microcomputers came along.

What you wish to have students do is the following:

- make their own data observations or experimental measurements;

- plot or graph the data in various ways (e.g. scatter plots, stem and leaf plots);

- summarize the data (mean, median, interquartile range, etc.);

- draw conclusions from the data (is it bimodal? are there outliers? etc.);

- transform the data (by logarithmic plots perhaps);

- smooth the data;

- compare different sets of data.

Readers will be able to add to this list.

The hand-held calculator and the microcomputer make it possible for students to do all the tasks listed above in reasonable amounts of time. Note that exploratory data analysis is not the same as doing statistics which is an interesting and important subject in its own right (Engel, Supporting Papers). Statistics involves such things as hypothesis testing, confidence intervals and analysis of variance, all expressed in a traditional mathematical framework involving lots of formulas. Exploratory data analysis provides less rigorous, less abstract subject matter which will be good preparation for a student for a course in statistics.

REFERENCES

Andrews, D.F. et al: 1972, Robust Estimates and Location, Princeton University Press.
Appel, K. and Haken, W. : 1976, 'The solution of the 4-color map problem', Sci. American (Oct.), 108-121.
Atkin, A.O.L. and Birch, B.J. (Eds.) : 1971, Computers in Number Theory, Academic Press.
Beker, H. and Piper, F. : 1982, Cipher Systems: the Protection of Communications, van Nostrand Reinhold.
Boieri, P. : 1984, 'An experiment of use of computers in basic mathematical courses', Proc 2nd Seminar on Mathematics in Engineering Education, Kassel.
Churchhouse, R.F. : 1969, 'An inexpensive information retrieval system', Proc SEAS XIV Conference.
Churchhouse, R.F. : 1973, 'Discoveries in number theory aided by computers', Bull. IMA, 9, 15-18.
Churchhouse, R.F. : 1980, 'Computer arithmetic and the failure of the associative law', Bull. IMA, 16, 210-214.

Churchhouse, R.F. : 1985, 'Computer arithmetic II: some computational
        anomalies and their consequences', Bull. IMA, 21, 70-73.
Churchhouse, R.F. and Herz, J.C. (Eds.) : 1968, Computers in Mathemati-
        cal Research, North Holland.
Davenport, J. : 1981, On the Integration of Algebraic Functions,
        Springer Lecture Notes in Computer Science 102.
Fatou, P. : 1919, 'Sur les equations fonctionelles', Bull. Soc. Math.
        France, 47, 161-271.
Fitch, F.B. : 1952, Symbolic Logic, An Introduction, Ronald Press,
        New York.
Fitch, J.P. : 1985, 'REDUCE', J. Symb. Comp. 1 (2), to appear.
Hardy, G.H. : 1916, The Integration of Functions of a Single Variable,
        Cambridge University Press.
Hodgson, B.R. and Poland, J.C. : 1983, 'Revamping the mathematical
        curriculum: the influence of computers', Notes Can. Math.
        Soc., 15 (8), 17-23.
Ingham, A.E. : 1932, The Distribution of Prime Numbers, Cambridge
        University Press.
Julia, G. : 1918, 'Memoire sur l'iteration des fonctions rationelles',
        J. Math. 8, 47-245.
Kessler, M.M. : 1965, 'The MIT technical information project', Physics
        Today (March), 28-36.
Lander, L.J. and Parkin, T.R. : 1967, 'A counter-example to Euler's sum
        of powers conjecture', Math. Comp., 21, 101-103.
Leech, J. (Ed.) : 1970, Computational Problems in Abstract Algebra,
        Pergamon Press.
Liouville, J. : 1833, J. de l'Ecole Polytechnique, 14, 124-193.
Mathematical Association of America : 1984, Panel Report on Discrete
        Mathematics (Preliminary version), Washington.
Pavelle, R., Fitch, J.P. and Rothstein, M. : 1981, 'Computer Algebra',
        Sci. American, 245, 102-113.
Pavelle, R. and Wang, P.S. : 1985, 'MACSYMA from F to G', J. Symb.
        Comp., 1, 69-100.
Ralston, A. : 1981, 'Computer science, mathematics and the under-
        graduate curricula in both', Amer. Math. Monthly, 88,
        472-485.
Ralston, A. and Young, G.S. : 1983, The Future of College Mathematics,
        Springer-Verlag.
Ramanujan, S. : 1927, Collected Papers, Cambridge University Press.
Slagle, J.R. : 1962, 'A heuristic program that solves symbolic integra-
        tion problems in freshman calculus', J.A.C.M., 10,
        507-520.
Tait, K. and Hughes, I.E. : 1984, 'Some experiences in using a
        computer-based learning system as an aid to self-teaching
        and self-assessment', Computer Educ., 8, 271-278.
Tall, D. : 1985, 'Understanding the calculus', Maths. Teaching, (March),
        49-53.
Tukey, J.W. : 1977, Exploratory Data Analysis, Addison Wesley.
Winkelmann, B. : 1984a, 'The impact of the computer on the teaching of
        analysis', Int. J. Math. Educ. Sci. Tech., 15, 675-689.

Winkelmann, B. : 1984b, 'Veranderungen von Zielsetzungen des
        Analysisunterrichts in Computerzeitalter', in W. Arlt and
        K. Haefner (Eds.) Herausforderung an Schule und Ausbildung,
        Springer-Verlag.
Zabusky, N.J. and Kruskal, M.D. : 1965, 'Interaction of "solitons" in a
        collisionless plasma and the recurrence of initial states',
        Phys. Rev. Lett., 15, 240-243.

List of Participants

Strasbourg, 25-30 March, 1985.

R. Allen, France.
F. Apery, France.
R. Biehler, F.R.G.
G. Bitter, U.S.A.
K. Bogart, U.S.A.
N. de Bruijn, The Netherlands.
H. Burkhardt, England.
D.E. Butler, England.
R. Churchhouse, Wales.
J.P. Conze, France.
B. Cornu, France.
J.H. Davenport, England.
C. Debieve, Belgium.
C. Delmez, Belgium.
E. Dubinsky, U.S.A.
C. Dupuis, France.
A. Engel, F.R.G.
R. Fraser, England.
K. Graf, F.R.G.
J. Guidy Wandja, Ivory Coast.
J. Hebenstreit, France.
B. Hodgson, Canada.
J.P. Houben, Belgium.
A.G. Howson, England.
J.P. Igot, France.
D. Jakubowicz, France.
J.P. Kahane, France.
L.H. Klingen, F.R.G.
S.K. Kivela, Finland.
C. Kuck, F.R.G.
K. Lane, U.S.A.
H. Lehning, France.
C.-K. Lim, Malaysia.
B. Mandelbrot, U.S.A.

A. Marzollo, Unesco.
M. Mascarello, Italy.
F. Massobrio, Italy.
H. Meissner, F.R.G.
B. Morin, France.
E. Muller, Canada.
H. Murakami, Japan.
M. Newman, Australia.
J. Niman, U.S.A.
A. Ollongren, The Netherlands.
J. Okon, U.S.A.
M. Otte, F.R.G.
F. Pluvinage, France.
H. Pollak, U.S.A.
A. Ralston, U.S.A.
D. Salinger, England.
R. Sargent, England.
D. Saunders, England.
B. Schmitt, France.
G. Schuyten, Belgium.
S. Seidman, U.S.A.
L.A. Steen, U.S.A.
J. Stern, France.
R. Strässer, F.R.G.
O. Takenouchi, Japan.
D.O. Tall, England.
M. Thorne, Wales.
J. Vanhamme, Belgium.
J. van Lint, The Netherlands.
R. Vezina, Canada.
B. West, U.S.A.
B. Winkelmann, F.R.G.
M.J. Winter, U.S.A.
M. Yamaguti, Japan.

SUPPORTING PAPERS

The following papers, submitted for consideration at the Strasbourg
meeting, are published in a supplementary volume by the IREM, 10 rue du
Général Zimmer, 67084 Strasbourg, France.

Contents List

1. ICMI Discussion Document.

General

2. Straesser, R., Remarks on the interplay of mathematics, computers
        and society.
3. Yamaguti, M., The influence of computers and informatics on
        mathematics.
4. Steen, L.A., Living with a new mathematical species.
5. Guidy-Wandja, J., L'informatique et l'enseignement dans les pays
        en voie de développement.
6. Klingen, L.H., Heutige und zukünftige didaktische Zustäude.
7. Graf, K.D., Informatics and mathematics in school education.
8. Monakhov, V.M., The information science and computing technique
        study influence upon school education.

Mathematics education and curricula

9. Thorne, M., Computers as a university mathematics teaching aid:
        towards a strategy.
10. Sargent, D.R., Computers as teaching aid.
11. Dubinski, E., Computer experience as an aid in learning
        mathematics concepts.
12. Kapadia, R., A new degree in Mathematical Studies.
13. Salinger, D.L., The effect of computers on the teaching of
        mathematics students.
14. Murakami, H. and Masoto, H., The progress of computers and
        mathematical education in Japan.
15. Davenport, J., The University of Bath syllabuses.
16. Butler, D., The M.E.I. schools project.
17. Seda, A.K., Computer science and the mathematics curriculum.
18. Burkhardt, H., Computer aware curricula: Ideas and realisation.

Some experiences

19. Winter, M.J., Using computers with undergraduate mathematics
        students in college algebra, elementary calculus, and
        teacher-training courses.
20. Allen, R., Nicolas, P. and Trilling, L., Learning geometry with
        the assistance of logic programming.
21. IREM Strasbourg, Experiences sur les apports de l'informatique à
        l'enseignement des mathématiques.

22. Bitter, G., Cameron, A. and Pitcher, S., First year results of the
        microcomputer assisted mathematics remediation project at
        Arizona State University.
23. Saunders, D.J., Computer animations in mathematics teaching at the
        Open University.
24. Mason, J.H., What happens when you switch off the machine.

Symbolic systems and algebra

25. de Bruijn, N.G., The Automath Mathematics checking project and its
        influence on teaching.
26. Lane, K.D., Symbolic manipulators and the teaching of college
        mathematics.
27. Stoutemyer, D.R., Using computer symbolic math for learning by
        discovery.
28. Ollongren, A., Formula manipulation in teaching perturbation
        methods.
29. Newman, M.F., Some software for teaching algebra.
30. Calmet, J., Introducing computer algebra to users and to students.
31. Lazard, D., Effet de l'informatique sur la rédaction et la pensée
        mathématique.
32. Stern, J., Réflexions sur certaines bases mathématiques de
        l'informatique.

Stochastics

33. Biehler, R., Interrelations between computers, statistics and
        teaching mathematics.
34. Engel, A., Algorithmic aspects of stochastics.
35. Conze, J.P., Les calculateurs et l'enseignement des probabilités.

Discrete and continuous mathematics

36. Harthong, J., Quelques éléments pour une théorie du continu.
37. Seidman, S.B. and Rice, M.D., A fundamental course in higher
        mathematics incorporating discrete and continuous themes.
38. Jenkyns, T.A. and Muller, E.R., Discrete mathematics – Two years
        experience with an introductory course.
39. Bogart, K.P., What should a discrete mathematics course be?
40. Hodgson, B.R., Muller, E.R., Poland, J., Taylor, P.D.,
        Introductory calculus in 1990.
41. Takenouchi, O., Computers in the beginner's course of the calculus.
42. Mascarello, M. and Scarafiotti, A.R., Computer experiments on
        mathematical analysis teaching at the Politecnico of Torino.
43. Winkelmann, B., The impact of the computer on the teaching of
        analysis.

Visualizations

44. Kivelä, S.K., The influence of computer graphics on the teaching
        of geometry.
45. Apery, F., Les surfaces peurent-elles être representées par
        l'ensemble des zéros d'un polynome à trois variables?
46. Tall, D., Visualizing calculus concepts using a computer.
47. West, B., Computer graphics as an essential research tool in the
        iteration of rational functions.
48. Artigue, M.,Gaitheron, V. and Isambert, E., L'étude graphique
        réhabilitée par l'ordinateur.
49. Yamaguti, M. and Masayoshi, H., On some multigrid finite
        difference schemes which describe everywhere non-
        differentiable functions.

MATHEMATICS AND THE COMPUTER REVOLUTION

M.F. Atiyah
Mathematical Institute, Oxford University, Oxford, England.

1 A HISTORICAL PERSPECTIVE

This Orwellian year of 1984 provides an inviting occasion for us to look to the past, present and future of mankind and in particular to consider the constantly changing relations between Science and Society. While George Orwell pin-pointed with great dramatic effect many of the political dangers of "double-think", the perversion of truth for political ends, he underestimated in other ways the enormous changes which Science had in store for us. The major problem we face today is of course the existence of atomic weapons and our capacity to destroy civilization, but even if this problem is solved many other challenges remain and prominent among these is the computer revolution.

It is now commonly acknowledged that we are firmly embarked on an economic and social revolution which will be comparable in scope and effect to the industrial revolution. There are here many significant analogies but also many important differences, notably in the speed of change. Whereas the industrial revolution is usually measured in centuries, the computer revolution is properly measured in decades. Since the human life-span has not fundamentally altered, the impact of the computer revolution will be faster and more acute in sociological terms, and coming to terms with it will be correspondingly more difficult.

Not being an economist or a sociologist I will leave it to others to elaborate on the obvious problems and likely developments in these areas. As a mathematician I am more concerned with another aspect of the computer revolution and one in which it differs fundamentally from its predecessor the industrial revolution. Whereas the eighteenth and nineteenth centuries witnessed the gradual replacement of manual labour by machines, the late twentieth-century is seeing the mechanization of intellectual activities. It is the brain rather than the hand that is now being made redundant. This means that the challenge which we face is of quite a different order and analogies with the past may therefore be misleading.

The intellectual challenge presented by the computer is, I believe, very far-reaching even if at the present time the problems are only just beginning to emerge. Moreover this challenge is certainly not restricted to mathematics, but will eventually penetrate into almost all aspects of human activity. For example, we are already seeing the introduction of "expert systems" into fields such as medicine and law, and the roles of doctors and judges as we now understand them are unlikely to survive unchanged into the next century. Science fiction in these directions has difficulty in keeping pace with fact.

Exciting though it is to speculate on the computerization of thought and knowledge in such fields I will, for two reasons, restrict myself to mathematics. The first and most important is that I am myself a mathematician and that I can speak about this area at first-hand and with some confidence. The temptation to pretend to expertise which one does not possess should be firmly avoided. The second reason for concentrating on mathematics is that, in the public eye at least, this subject tends to be naturally, though not always correctly, associated with computers.

It is of course true that, in its early days, computer science grew up alongside mathematics and that famous mathematicians such as Turing and von Neumann were amongst its pioneers. Moreover, of the traditional basic sciences, mathematics is still the closest in spirit to computer science. In fact, it is sometimes asserted, with dry humour, that computer science is the Cuckoo of the mathematical nest with all the unpleasant overtones which that suggests.

In this world of education, mathematics and computer science still go hand in hand even if the relationship is now an uneasy one. In universities, Computing and Mathematics are frequently found together and, at the school level, computing is almost exclusively in the hands of mathematics teachers.

For all these reasons it seems to me that mathematicians have a responsibility to explain, to society at large, intellectual challenges and dangers presented by the computer revolution. This is what I hope to do today. As I have already mentioned I shall restrict myself to describing the impact of computers on mathematics itself, but at a fundamental level I believe that many of the things I shall say will have some relevance to other fields of intellectual study, though I will leave it to each of you to decide how far my remarks pertain to your own discipline or field of interest.

Finally, I should issue a disclaimer. Some of my mathematical colleagues have much more direct experience of computing than I have. In fact, at the technical level I am barely a novice, but I hope that, at a higher level, I am aware of what is happening and that my perception is not too far off the mark.

2 MATHEMATICS AND THEORETICAL COMPUTER SCIENCE

It might be helpful if I began by describing the role which mathematics has played, and continues to play, in the development of the theoretical aspects of computers. Not surprisingly, those parts of mathematics which have been relevant in this respect have themselves received an enormous stimulus in return. On the one hand, this has been beneficial in many ways by suggesting fruitful lines of research, but the sheer scale of the computer field brings dangers in its wake, and I shall return to this later.

Historically it was mathematical logic which provided the theoretical basis for computers. Here, and throughout my lecture, I am referring to the 'software' side of computers, concerned with the development and use of suitable languages rather than the 'hardware' side which is concerned with their physical design and construction. Of course, it is the hardware development − the rise of the minute silicon chip − which has produced the revolution, but this in its turn only re-emphasizes the need for greater sophistication in the language so as to fully exploit the hardware potentialities.

Mathematicians have always been concerned with the notion of "proof", the rigorous deduction of various conclusions from given assumptions, and in the first half of the twentieth-century this notion was subjected to extremely careful analysis. In particular there emerged the notion of a "constructive" proof, where the desired conclusion could be arrived at after a finite number of definite steps. The famous "Turing machine" was a hypothetical ideal machine which could carry out such constructive proofs, and the early computers were in essence its physical realization.

Computers have to be given precise commands and mathematical logic provides the theoretical framework in which such commands can be formulated. Moreover as computers become more and more powerful, so the languages they use become increasingly sophisticated, and the problems of possible errors loom much larger. The errors I refer to are not of course machine errors − a machine can do no wrong − but of human errors in issuing the right commands or in translating into the computer language. Here again mathematical ideas of proof become important − how to prove that a given set of computer instructions are correct.

This very briefly is why mathematical logic is related to theoretical computer science and why students trained in this most abstract of mathematical disciplines find a ready demand for their talents in the computer field.

Closely related to the notion of constructive proof is that of an "algorithm" which in mathematical parlance is the term used for a definite procedure for solving a problem. For example an explicit formula for solving an equation is an especially simple algorithm. If a mathematician wants to use a computer to solve a problem he needs to give the computer an algorithm. Now algorithms can be fast or slow, measured in computer time, and there is clearly a great advantage in devising fast algorithms. Thus the development of computers has stimulated a whole new branch of mathematics, complexity theory, which

is essentially concerned with understanding how "complex" an algorithm
is and which roughly corresponds to how long a computer will take to
give an answer.

Proof theory and complexity are just two examples of the sort of
mathematics which has been stimulated or created by the needs of the
computer. In general the sort of mathematics involved is quite
different from that required by the applications to physical science.
Because computers are based on the on/off switch of electrical circuits
they involve discrete mathematics exemplified by algebra, whereas,
since Newton's time, physical science has been based largely on the
application of calculus to the study of continuously varying phenomena.
This has led some people to argue that the traditional approach to
teaching mathematics, with its heavy emphasis on calculus must, in the
age of the computer, be drastically modified.

3   COMPUTERS AS AN AID TO MATHEMATICAL RESEARCH
Having described the way in which mathematics has helped
the development of computer science, let me now consider the flow in
the opposite direction. In what ways has the advent of computers
assisted and altered mathematical research?

The first and most obvious use of computers has been simply as "number
crunchers". High-speed machines are excellently adapted to carrying
out very large numbers of repetitive calculations, so that explicit
numerical answers can rapidly be provided for problems which would
otherwise have been too complicated to handle. This use of computers
has had a dramatic effect on all of applied mathematics and it has
significantly altered our conception of what is a satisfactory
solution of a mathematical problem. In pre-computer days mathematicians would work hard to cast the solution of a problem into some
elegant algebraic form, involving familiar objects such as algebraic
and trigonometric expressions. Nowadays a problem in applied
mathematics is regarded as satisfactorily solved if one can find an
algorithm to feed into a computer which will generate all the
numerical values one is interested in.

Not all of mathematics however is concerned with numbers. Algebra for
instance deals with symbolic expressions which may or may not stand
for unknown numbers. For example, an expression in mathematical logic
does not stand for anything numerical. The manipulation of complicated
symbolic expressions can also be performed on computers and there are
areas of mathematics where this has already been applied very
successfully. For example, the determination of all finite simple
groups, the building blocks of abstract symmetries, was greatly
assisted by the use of powerful computers. With the increased
availability of micro-computers and the greater computer expertise
among the younger generation of mathematicians it seems certain that
these symbolic uses of computers will greatly increase.

In Mathematics, as in the Natural Sciences, there are several stages involved in a discovery, and formal proof is only the last. The earliest stage consists in the identification of significant facts, their arrangement into meaningful patterns and the plausible extraction of some law or formula. Next comes the process of testing this proposed formula against new experimental facts, and only then does one consider the question of proof.

In all the earlier stages computers can play a role, particularly when large or complex systems are being considered. For example, in Number Theory interesting questions may involve very large prime numbers, and some of the deepest conjectures being studied at the present time have been based on extensive computer calculations. In the same way problems in differential equations which involve the evolution of some system (e.g. the flow of a liquid) for a very long time have been enormously influenced by experimental facts discovered on computers.

One advantage of present day computers which is only just beginning to be fully appreciated by mathematicians is their ability to display information graphically (and even in colour). For many complicated mathematical problems involving geometrical features, this provides an extremely effective new tool with which to explore phenomena.

To sum up therefore the computer is proving of great practical assistance to mathematicians at all stages of their work, but perhaps most significantly in the exploratory or experimental stage. Great mathematicians of the past such as Euler or Gauss carried out large numbers of tedious calculations by hand in order to provide themselves with the raw material from which they could then guess some general law, or discover some remarkable pattern. As mathematics delves further and we become more ambitious the raw material becomes correspondingly much more messy and complicated. The computer can help us to sift this material and to point the way to further progress and understanding.

4   THE INTELLECTUAL DANGERS

Few scientific advances are unmixed blessings and the computer is no exception. Having enumerated the many benefits which mathematicians, amongst others, can derive from the advent of the computer I would like now to draw attention to some possible dangers that lie ahead. Let me begin with the most central and insidious problem which is essentially the challenge the computer presents to the human intellect. Will mathematics continue as one of the highest forms of human endeavour or will it gradually be taken over by the computer? Who will remain in charge of mathematics and what are its criteria to be?

To illustrate the dangers I have in mind, let us consider an event that has already taken place, namely the solution by computer of a famous outstanding mathematical problem. I refer to the 4-colour

theorem which says roughly that four colours suffice to colour any
conceivable map of the world, the requirement being that adjacent
countries must be coloured differently. This problem which dates from
the last century was recently solved by a proof which involved a
computer check of hundreds of different cases. On the one hand, this
was a great triumph, the solution of a hard problem; on the other
hand from an aesthetic point of view, it is extremely disappointing,
and no new insights are derived from the proof.

Is this to be the way of the future? Will more and more problems be
solved by brute force? If this is indeed what is in store for us
should we be concerned at the decline of human intellectual activity
this represents, or is that simply an archaic view-point which must
give way before the forces of "progress"?

To answer such philosophical questions we must be bold and ask ourselves what is the nature and purpose of mathematical and scientific
activity. The usual answer is that Science is man's attempt to
understand, and perhaps eventually control, the physical world, but
this leaves us with the difficult notion of "understanding". Can we
be said to "understand" the proof of the 4-colour theorem? I doubt it.

For those who feel that "understanding" is too subjective and restric-
tive a term, the more limited goal of "description" may be preferred.
Certainly I can describe the proof of the 4-colour theorem, though my
description entails saying "the computer checked the following facts".

Such a "descriptive" attitude to mathematics could live happily with a
gradual take-over by the computer, but I believe that this would lead
to the atrophy of mathematics even measured by these modest
"descriptive" standards. Mathematics is really an Art - it is the art
of avoiding brute-force calculation by developing concepts and techniques which enable one to travel more lightly. Give a mathematician
an infinitely powerful machine for doing calculations and you deprive
him of his inner driving force. It is at least arguable, though
somewhat far-fetched, that if computers had been available in say the
fifteenth century, mathematics now would be a pale shadow of itself.

5   ECONOMIC DANGERS
In addition to the subtle and intangible intellectual
threat posed by computers there are much more obvious and practical
dangers due to the tremendous economic importance of computers to
society at large. Inevitably there will be vast financial pressures
which will tend to push mathematics into new directions related to
computing. Broadly speaking, more emphasis will be put on discrete
mathematics as opposed to Calculus, which is concerned with continuous
phenomena. No doubt, some of this pressure will be healthy and will
stimulate and generate exciting new branches of mathematics, but the
scale and tempo of the computer revolution are such that there is a
real danger of the great classical tradition of mathematics being
swamped.

Superficially, at first sight, discrete mathematics, which deals only with finite quantities and processes is easier and simpler than Calculus which deals with the infinite in various forms. However, it is one of the greatest triumphs of mathematics that infinity has been tamed and put to use, so that Calculus is a tool of enormous power and elegance which has no serious rival or counterpart at the purely finite level. In fact, many important results of a discrete nature are best proved by the use of Calculus.

Until now the central position of Calculus, the Analysis of the infinite, has been unassailable not only in Pure Mathematics, but even more as a foundation for the application of mathematics to the whole of Science and Engineering. Courses in Calculus have provided the bed-rock of University education in the Mathematical Sciences. In recent years however this position has been called into question and there is an increasing call to reduce the role of Calculus in scientific education and replace it with the kind of discrete mathematics more relevant to Computer Studies. To some extent this has already happened and it represents a necessary response to changing conditions, but I foresee pressures for much more radical changes which might be very damaging but would be difficult to resist.

It is possible that I am being unduly pessimistic on this score and certainly the dichotomy between discrete and continuous mathematics is not as sharp as I have implied. Traditionally we think of using finer and finer discrete quantities to approximate a continuous quantity in the way a continuous curve can be approximated by a large number of straight line segments. However, this procedure can equally well be reversed and continuous quantities can be regarded as approximations to discrete ones, provided the step size is sufficiently small. Thus we can use our knowledge (derived from Calculus) for the length of a circle to get an approximation for the length of a regular polygon with a large number of sides. In this way, as computers become more and more powerful and the numbers they deal with become larger and larger (or the time span for a single operation becomes shorter and shorter) so Calculus may again come into its own.

6 EDUCATIONAL DANGERS

As we all know, the present economic scene is of wide-spread decline of traditional industries and the simultaneous growth of computer related industries. This is the economic side of the revolution. Naturally, this means that the best employment opportunities are linked to computers and this is altering the attitudes and expectations of all the younger generation. In schools and universities traditional studies are having to compete with the excitement and attraction of the computer, but mathematics, as the closest of the older discipline, is inevitably in the front line. This is having an effect at several levels. In the first place the pressure falls on the mathematics teachers in schools. Increasingly, they are having to take on computer studies as an additional responsibility and this means that mathematics teaching as such is

suffering. Our educational institutions, for organizational and
human reasons, can only change slowly and the sheer speed of the
computer revolution is going to put them under very severe strain.

As far as students are concerned mathematics is going to be affected
in two different ways. For the abler student who might have gone on
to creative work in the higher reaches of mathematics, there is now
the attractive alternative of entering a field which is in an
explosive stage of its development and where the opportunities to make
your mark are much greater. This means that the great creative minds
of the past such as Newton, Gauss or Riemann might in future gravitate
towards computer science rather than mathematics. For a subject so
entirely dependent on brain-power this would be the greatest disaster
of all. One has to hope that mathematics, by its power and beauty,
will still attract intellects of quality in the future and that not
all of them will be seduced by the computer.

For students of mathematics at a lower level, there are other dangers.
At their most elementary these are simply that the wide-spread use of
computers, or even sophisticated calculators, will lead to the view
that arithmetic is no longer a necessary skill to acquire. Why learn
your multiplication tables when, at the push of a button, the answer
will appear on your screen? Such attitudes are already with us and
there is much educational debate on say the advantages and dangers of
having calculators in primary schools. As computers become even
cheaper and more powerful, they will flood into our schools and
mathematics at all levels will constantly have to justify itself.

The enlightened response to these philistine attacks on mathematics is
to say that, even when all the work can be done by pushing a button,
you have to teach children which button to push. At the most basic
level they have to know when to push the addition sign and when to
push the multiplication sign. This means that there has to be more
emphasis on understanding the processes involved and less on the
performance of routine calculations. Properly interpreted this can be
regarded as an educational advantage in which drudgery is removed and
appreciation is enhanced. However, life is not quite so simple and
any over-reliance on machines can lead to the atrophy of the human
faculties involved, much in the way the motor car has undermined the
capacity of people to use their legs. Perhaps the sort of reaction
which has made jogging so popular in recent years will in due course
make exercise in mental arithmetic a form of mental therapy!

7   CONCLUSION
I have tried to draw attention to the challenges and
dangers which the rise of computers presents to mathematics. I am
sorry if the picture I have been depicting appears too negative. It
is easy to see the benefits which mathematics may derive from its
association with computing, and so I have not thought it worthwhile
to emphasize these at length. The dangers I think are more subtle and
not so well recognized so it seemed appropriate to dwell on them in

greater detail. To recognize possible dangers is to be fore-armed,
and one can hope to prevent the worst from actually happening.
Perhaps I can end on this note by recalling that George Orwell did
not view his book on 1984 as a prediction but as a warning,
deliberately exaggerated for dramatic effect, of what might happen
if we were not careful.

(This paper is based on a lecture given by Professor Atiyah at the
Locarno Conference (May, 1984) on the theme "1984: comincia il futuro".
The conference was organised by the Dipartimento Pubblica Educazione
Cantone Ticino and the journal Nuova Civilità della Macchine (Bologna).
The Italian translation of the lecture appeared in Vol 2 no. 3 (1984)
of this journal. ICMI is grateful to Professor Barone, editor of
Nuova Civilità della Macchine, for granting us permission to include
Sir Michael's paper in this volume.)

LIVING WITH A NEW MATHEMATICAL SPECIES

Lynn Arthur Steen
St. Olaf College, Northfield, Minn., 55077, U.S.A.

Computers are both the creature and the creator of mathematics. They are, in the apt phrase of Seymour Papert, "mathematics-speaking beings". More recently J. David Bolter in his stimulating book Turing's Man [4] calls computers "embodied mathematics". Computers shape and enhance the power of mathematics, while mathematics shapes and enhances the power of computers. Each forces the other to grow and change, creating, in Thomas Kuhn's language, a new mathematical paradigm.

Until recently, mathematics was a strictly human endeavor. But suddenly, in a brief instant on the time scale of mathematics, a new species has entered the mathematical ecosystem. Computers speak mathematics, but in a dialect that is difficult for some humans to understand. Their number systems are finite rather than infinite; their addition is not commutative; and they don't really understand "zero", not to speak of "infinity". Nonetheless, they do embody mathematics.

The core of mathematics is changing under the ecological onslaught of mathematics-speaking computers. New specialties in computational complexity, theory of algorithms, graph theory, and formal logic attest to the impact that computing is having on mathematical research. As Arthur Jaffe has argued so well (in [12]), the computer revolution is a mathematical revolution.

New Mathematics for a New Age

Computers are discrete, finite machines. Unlike a Turing machine with an infinite tape, real machines have limits of both time and space. Theirs is not an idealistic Platonic mathematics, but a mathematics of limited resources. The goal is not just to get a result, but to get the best result for the least effort. Optimization, efficiency, speed, productivity--these are essential objectives of modern computer mathematics.

Computers are also logic machines. They embody the fundamental engine of mathematics--rigorous propositional calculus. The first celebrated computer proof was that of the four-color theorem: the computer served there as a sophisticated accountant, checking out thousands of cases of reductions. Despite philosophical alarms that computer-based proofs change mathematics from an a priori to a contingent, fallible subject (see, e.g., [27]), careful analysis reveals that nothing much has really changed. The human practice of mathematics has always been fallible; now it has a partner in fallibility.

Recent work on the mysterious Feigenbaum constant reveals just how far this evolution has progressed in just eight years: computer-assisted investigations of families of periodic maps suggested the presence of a mysterious universal limit, apparently independent of the particular family of maps. Subsequent theoretical investigations led to proofs that are true hybrids of classical analysis and computer programming [8], showing that computer-assisted proofs are possible not just in graph theory, but also in functional analysis.

Computers are also computing machines. By absorbing, transforming, and summarizing massive quantities of data, computers can simulate reality. No longer need the scientist build an elaborate wind tunnel or a scale model refinery in order to test engineering designs. Wherever basic science is well understood, computer models can emulate physical processes by carrying out instead the process implied by mathematical equations. Whereas mathematical models used to be primarily tools used by theoretical scientists to formulate general theories, now they are practical tools of enormous value in the everyday world of engineering and economics.

It has been fifty years since Alan Turing developed his seminal scheme of computability [26], in which he argued that machines could do whatever humans might hope to do. In abstract terms, what he proposed was a universal machine of mathematics (see [11] for details). It took two decades of engineering effort to turn Turing's abstract ideas into productive real machines. During that same period abstract mathematics flourished, led by Bourbaki, symbolized by the "generalized abstract nonsense" of category theory. But with abstraction came power, with rigor came certainty. Once real computers emerged, the complexity of programs quickly overwhelmed the informal techniques of backyard programmers. Formal methods became de rigueur; even the once-maligned category theory is now enlisted to represent finite automata and recursive functions (see, e.g., [2]). Once again, as happened before with physics, mathematics became more efficacious by becoming more abstract.

The Core of the Curriculum
Twenty years ago in the United States the Committee on the Undergraduate Program in Mathematics (CUPM) issued a series of reports that led to a gradual standardization of curricula among undergraduate mathematics departments [5]. Shortly thereafter, in 1971, Garrett Birkhoff and J. Barkley Rosser presented papers at a meeting of the Mathematical Association of America concerning predictions for undergraduate mathematics in 1984. Birkhoff urged increased emphasis on modelling, numerical algebra, scientific computing, and discrete mathematics. He also advocated increased use of computer methods in

pure mathematics: "Far from muddying the limpid waters of clear
mathematical thinking, [computers] make them more transparent by
filtering out most of the messy drudgery which would otherwise
accompany the working out of specific illustrations." [3, p. 651]
Rosser emphasized many of the same points, and warned of impending
disaster to undergraduate mathematics if their advice went unheeded:
"Unless we revise [mathematics courses] so as to embody much use of
computers, most of the clientele for these courses will instead be
taking computer courses in 1984." [21, p. 639]

In the decade since these words were written, U.S. undergraduate and
graduate degrees in mathematics have declined by 50%. The clientele
for traditional mathematics has indeed migrated to computer science,
and the former CUPM consensus is all but shattered. Five years ago
CUPM issued a new report, this one on the Undergraduate Program in
Mathematical Sciences [6]. Beyond calculus and linear algebra, they
could agree on no specific content for the core of a mathematics major:
"There is no longer a common body of pure mathematical information that
every [mathematics major] should know."

The symbol of reformation has become discrete mathematics. Several
years ago Anthony Ralston argued forcefully the need for change before
both the mathematics community [17] and the computer science community
[18]. Discrete mathematics, in Ralston's view, is the central link
between the fields. The advocacy of discrete mathematics rapidly
became quite vigorous (see, e.g., [19] and [24]), and the Sloan
Foundation funded experimental curricula at six institutions to
encourage development of discrete-based alternatives to standard
freshman calculus.

The need for consensus on the contents of undergraduate mathematics is
perhaps the most important issue facing American college and university
mathematics departments. On the one hand departments that have a
strong traditional major often fail to provide their students with the
robust background required to survive the evolutionary turmoil in the
mathematical sciences. Like the Giant Panda, these departments depend
for survival on a dwindling supply of bamboo--strong students
interested in pure mathematics. On the other hand, departments
offering flabby composite majors run a different risk: by avoiding
advanced, abstract requirements, they often misrepresent the true
source of mathematical knowledge and power. Like zoo-bred animals
unable to forage in the wild, students who have never been required to
master a deep theorem are ill-equipped to master the significant
theoretical complications of real-world computing and mathematics.

Computer Literacy
Mathematical scientists at American institutions of higher
education are responsible not only for the technical training of future
scientists and engineers, but also for the technological literacy of
laymen--of future lawyers, politicians, doctors, educators, and clergy.
Public demand that college graduates be prepared to live and work in a

computer age has caused many institutions to introduce requirements in
quantitative or computer literacy.

In 1981 the Alfred P. Sloan foundation initiated curricular exploration
of "the new liberal arts", the role of applied mathematical and
computer sciences in the education of students outside technical
fields. "The ability to cast one's thoughts in a form that makes
possible mathematical manipulation and to perform that manipulation ...
[has] become essential in higher education, and above all in liberal
education." [14, p. 6] Others echoed this call for reform of liberal
education. David Saxon, President of the University of California
wrote in a Science editorial that liberal education "will continue to
be a failed idea as along as our students are shut out from, or only
superficially acquainted with, knowledge of the kinds of questions
science can answer and those it cannot." [22]

Too often these days the general public views computer literacy as a
modern substitute for mathematical knowledge. Unfortunately, this
often leads students to superficial courses that emphasize vocabulary
and experiences over concepts and principles. The advocates of
computer literacy conjure images of an electronic society dominated by
the information industries. Their slogan of "literacy" echoes
traditional educational values, conferring the aura but not the logic
of legitimacy.

Typical courses in computer literacy are filled with ephemeral details
whose intellectual life will barely survive the students' school years.
These courses contain neither a Shakespeare nor a Newton, neither a
Faulkner nor a Darwin; they convey no fundamental principles nor
enduring truths. Computer literacy is more like driver education than
like calculus. It teaches students the prevailing rules of the road
concerning computers, but does not leave them well prepared for a
lifetime of work in the information age.

Algorithms and data structures are to computer science what functions
and matrices are to mathematics. As much of the traditional
mathematics curriculum is devoted to elementary functions and matrices,
so beginning courses in computing--by whatever name--should stress
standard algorithms and typical data structures. As early as students
study linear equations they could also learn about stacks and queues;
when they move on to conic sections and quadratic equations, they could
in a parallel course investigate linked lists and binary trees.

Computer languages can (and should) be studied for the concepts they
represent--procedures in Pascal, recursion and lists for Lisp--rather
than for the syntactic details of semicolons and line numbers. They
should not be undersold as mere technical devices for encoding problems
for a dumb machine, nor oversold as exemplars of a new form of
literacy. Computer languages are not modern equivalents of Latin or
French; they do not deal in nuance and emotion, nor are they capable
of persuasion, conviction, or humor. Although computer languages do
represent a new and powerful way to think about problems, they are not
a new form of literacy.

Computer Science
In the United States, computer science programs cover a
broad and varied spectrum, from business-oriented data processing
curricula, through management information science, to theoretical
computer science. All of these intersect with the mathematics
curriculum, each in different ways.

Recently Mary Shaw of Carnegie Mellon University put together a
composite report on the undergraduate computer science curriculum.
This report is quite forceful about the contribution mathematics makes
to the study of computer science: "The most important contribution a
mathematics curriculum can make to computer science is the one least
likely to be encapsulated as an individual course: a deep appreciation
of the modes of thought that characterize mathematics." [23. p. 55]

The converse is equally true: one of the more important contributions
that computer science can make to the study of mathematics is to
develop in students an appreciation for the power of abstract methods
when applied to concrete situations. Students of traditional
mathematics used to study a subject called "Real and Abstract
Analysis"; students of computer science now can take a course titled
"Real and Abstract Machines". In the former "new math", as well as in
modern algebra, students learned about relations, abstract versions of
functions; today business students study "relational data structures"
in their computer courses, and advertisers tout "fully relational" as
the latest innovation in business software.

An interesting and pedagogically attractive example of the power of
abstraction made concrete can be seen in the popular electronic
spreadsheets that are marketed under such trade names as Lotus and
VisiCalc. Originally designed for accounting, they can as well emulate
cellular automata or the Ising model for ferromagnetic materials [10].
They can also be "programmed" to carry out most standard mathematical
algorithms--the Euclidean algorithm, the simplex method, Euler's method
for solving differential equations [1]. An electronic spreadsheet--the
archetype of applied computing--is a structured form for recursive
procedures--the fundamental tool of algorithmic mathematics. It is a
realization of abstract mathematics, and carries with it much of the
power and versatility of mathematics.

Computers in the Classroom
Just as the introduction of calculators upset the
comfortable pattern of primary school arithmetic, so the spread of
computers will upset the traditions of secondary and tertiary
mathematics. This year long division is passe; next year integration
will be under attack.

The impact of computing on secondary school mathematics has been the
subject of many recent discussions in the United States. Jim Fey,
coordinator of two of the most recent assessments ([7], [9]), described
these efforts as "an unequivocal dissent from the spirit and substance

of efforts to improve school mathematics that seek broad agreement on
conservative curricula." [9, p. viii] Teachers in tune with the
computer age seek change in both curriculum and pedagogy. But the
inertia of the system remains high. For example, the recent
International Assessment of Mathematics documented that in the United
States calculators are never permitted in one-third of the 8th grade
classes, and rarely used in all but 5% of the rest [25, p. 18].

Lap size computers are now common--they cost about as much as ten
textbooks, but take up only the space of one. Herb Wilf argues (in
[28]) that it is only a matter of time before students will carry with
them a device to perform all the algorithms of undergraduate
mathematics. Richard Rand, in a survey [20] of applied research based
on symbolic algebra agrees: "It will not be long before computer
algebra is as common to engineering students as the now obsolete slide
rule once was."

Widespread use of computers that do school mathematics will challenge
standard educational practice. For the most part, computers reinforce
the student´s desire for correct answers. In the past, their school
uses have primarily extended the older "teaching machines": programmed
drill with pre-determined branches for all possible responses. But the
recent linking of symbolic algebra programs with so-called "expert
systems" into sophisticated "intelligent tutors" has produced a rich
new territory for imaginative computer-assisted pedagogy that advocates
claim can rescue mathematics teaching from the morass of rules and
template-driven tests.

It is commonplace now to debate the wisdom of teaching skills (such as
differentiation) that computers can do as well or better than humans.
Is it really worth spending one month of every year teaching half of a
country´s 18 year old students how to imitate a computer? What is not
yet so common is to examine critically the effect of applying to
mathematics pedagogy computer systems that are themselves only capable
of following rules or matching templates. Is it wise to devise
sophisticated computer systems to teach efficiently precisely those
skills that computers can do better than humans, particularly those
skills that make the computer tutor possible? In other words, since
computers can now do the calculations of algebra and calculus, should
we use this power to reduce the curricular emphasis on calculations or
to make the teaching of these calculations more efficient? This is a
new question, with a very old answer.

Let Us Teach Guessing
35 years ago George Polya wrote a brief paper with the
memorable title "Let Us Teach Guessing" [16]. It is not differentiation that our students need to learn, but the art of guessing. A
month spent learning the rules of differentiation reinforces a
student´s ability to learn (and live by) the rules. In contrast, time
spent making conjectures about derivatives will teach a student
something about the art of mathematics and the science of order.

With the aid of the mathematics-speaking computer, students can for the first time learn college mathematics by discovery. This is an opportunity for pedagogy that mathematics educators cannot afford to pass up. Mathematics is, after all, the science of order and pattern, not just a mechanism for grinding out formulas. Students discovering mathematics gain insight into the discovery of pattern, and slowly build confidence in their own ability to understand mathematics. Formerly, only students of sufficient genius to forge ahead on their own could have the experience of discovery. Now with computers as an aid, the majority of students can experience for themselves the joy of discovery.

Metaphors for Mathematics
Two metaphors from science are useful for understanding the relation between computer science, mathematics, and education. Cosmologists long debated two theories for the origin of the universe-- the Big Bang theory, and the theory of Continuous Creation. Current evidence tilts the cosmology debate in favor of the Big Bang. Unfortunately, this is all too often the public image of mathematics as well, even though in mathematics the evidence favors Continuous Creation.

The impact of computer science on mathematics and of mathematics on computer science is the most powerful evidence available to beginning students that mathematics is not just the product of an original Euclidean big bang, but is continually created in response to challenges both internal and external. Students today, even beginning students, can learn things that were simply not known 20 years ago. We must not only teach new mathematics and new computer science, but we must teach as well the fact that this mathematics and computer science is new. That's a very important lesson for laymen to learn.

The other apt metaphor for mathematics comes from the history of the theory of evolution. Prior to Darwin, the educated public believed that forms of life were static, just as the educated public of today assumes that the forms of mathematics are static, laid down by Euclid, Newton and Einstein. Students learning mathematics from contemporary textbooks are like the pupils of Linnaeus, the great eighteenth century Swedish botanist: they see a static, pre-Darwinian discipline that is neither growing nor evolving.   Learning mathematics for most students is an exercise in classification and memorization, in labelling notations, definitions, theorems, and techniques that are laid out in textbooks as so much flora in a wonderous if somewhat abstract Platonic universe.

Students rarely realize that mathematics continually evolves in response to both internal and external pressures. Notations change; conjectures emerge; theorems are proved; counterexamples are discovered. Indeed, the passion for intellectual order combined with the pressure of new problems--especially those posed by the computer-- force researchers to continually create new mathematics and archive old theories.

The practice of computing and the theory of computer science combine to change mathematics in ways that are highly visible and attractive to students. This continual change reveals to students and laymen the living character of mathematics, restoring to the educated public some of what the experts have always known—that mathematics is a living, evolving component of human culture.

REFERENCES

1. Arganbright, Dean E. Mathematical Applications of Electronic
        Spreadsheets. McGraw-Hill, 1985.
2. Beckman, Frank S. Mathematical Foundations of Programming. The
        Systems Programming Series, Addison Wesley, 1984.
3. Birkhoff, Garrett. "The Impact of Computers on Undergraduate
        Mathematics Education in 1984." American Mathematical
        Monthly 79 (1972) 648-657.
4. Bolter, J. David. Turing's Man: Western Culture in the Computer
        Age. University of North Carolina Press, Chapel Hill,
        1984.
5. Committee on the Undergraduate Program in Mathematics. A General
        Curriculum in Mathematics for Colleges. Mathematical
        Association of America, 1965.
6. Committee on the Undergraduate Program in Mathematics.
        Recommendations for a General Mathematical Sciences
        Program. Mathematical Association of America, 1980.
7. Corbitt, Mary Kay, and Fey, James T. (Eds.). "The Impact of
        Computing Technology on School Mathematics: Report of an
        NCTM Conference." National Council of Teachers of
        Mathematics, 1985.
8. Eckmann, J.-P., Koch, H., and Wittwer, P. "A computer-assisted
        proof of universality for area-preserving maps." Memoirs of
        the American Mathematical Society, Vol 47, No. 289 (Jan.
        1984).
9. Fey, James T., et al. (Eds.). Computing and Mathematics: The
        Impact on Secondary School Curricula. National Council of
        Teachers of Mathematics, 1984.
10. Hayes, Brian. "Computer Recreations." Scientific American
        (October 1983) 22-36.
11. Hodges, Andrew. Alan Turing: The Enigma. Simon and Schuster,
        1983.
12. Jaffe, Arthur. "Ordering the Universe: The Role of Mathematics."
        In Renewing U. S. Mathematics, National Academy Press,
        Washington, D. C. 1984.
13. Kemeny, John G. "Finite Mathematics—Then and Now." In Ralston,
        Anthony and Young, Gail S. The Future of College
        Mathematics. Springer-Verlag, 1983, pp. 201-208.
14. Koerner, James D., ed. The New Liberal Arts: An Exchange of
        Views. Alfred P. Sloan Foundation, 1981.
15. Lewis, Harry R. and Papadimitriou. Elements of the Theory of
        Computation. Prentice-Hall. 1981.
16. Polya, George. "Let Us Teach Guessing." Etudes de Philosophie des

Sciences. Neuchatel: Griffon, 1950, pp. 147–154; reprinted in George Polya: Collected Papers. Vol. IV, MIT Press, 1984, pp. 504–121.

17. Ralston, Anthony. "Computer Science, Mathematics, and the Undergraduate Curricula in Both." American Mathematical Monthly 88 (1981) 472–485.

18. Ralston, Anthony and Shaw, Mary. "Curriculum '78: Is Computer Science Really that Unmathematical?" Communications of the ACM 23 (Feb. 1980) 67–70.

19. Ralston, Anthony and Young, Gail S. The Future of College Mathematics. Springer Verlag, 1983.

20. Rand, Richard H. Computer Algebra in Applied Mathematics: An Introduction to MACSYMA. Research Notes in Mathematics No. 94, Pitman Publ., 1984.

21. Rosser, J. Barkley. "Mathematics Courses in 1984." American Mathematical Monthly 79 (1972) 635–648.

22. Saxon, David S. "Liberal Education in a Technological Age." Science 218 (26 Nov 1982) 845.

23. Shaw, Mary (Ed.) The Carnegie–Mellon Curriculum for Undergraduate Computer Science. Springer Verlag, 1984.

24. Steen, Lynn Arthur. 1 + 1 = 0: New Math for a New Age. Science 225 (7 Sept. 1984) 981.

25. Travers, Kenneth, et. al. Second Study of Mathematics: United States Summary Report. University of Illinois, September 1984.

26. Turing, Alan M. "On Computable Numbers, with an Application to the Entscheidungsproblem." Proc. London Math. Soc. 2nd Ser., 42 (1936) 230–265.

27. Tymoczko, Thomas. "The Four Color Problem and its Philosophical Significance." Journal of Philosophy 76:2 (1979) 57–85.

28. Wilf, Herbert. "The Disk with the College Education." American Mathematical Monthly 89 (1982) 4–8.

CHECKING MATHEMATICS WITH THE AID OF A COMPUTER

N.G. de Bruijn, Department of Mathematics and Computing
Science, Technological University Eindhoven,
The Netherlands.

0.   Computers influence mathematics in many ways. This paper is
devoted to one of these influences: the fact that we can explain
mathematics to a computer. In this process we may learn about how to
organize mathematics and how to teach some of its aspects.

At the Technological University Eindhoven (Eindhoven, the Netherlands)
the project Automath was developed from 1967 onwards, with various kinds
of activities at the interfaces of logic, mathematics, computer science,
language and mathematical education. Right from the start, it was
directed towards the presentation of formalized knowledge to a computer,
in a very general language, with quite a strong emphasis on doing things
the way humans do. One might say that the project is a modern version
of "Leibniz's dream" of making a language for all scientific discussion
in such a way that all reasoning can be represented by a kind of
algebraic manipulation.

The basic idea of Automath is that the human being presents any kind of
discourse, however long it may be, to a machine, and that the machine
convinces itself that everything is sound. All this is intended to be
effectively carried out on a large scale, and not just "in principle".

This paper does not intend to describe the Automath system in any
detail, but rather to explain a number of goals, achievements and
characteristics that may have a bearing on the subject of the ICMI
discussion on the influence of computers and informatics on mathematics
and its teaching.

The paper is definitely not trying to sell Automath as a subject to be
taught to all students in standard mathematics curricula. The claim is
much more modest: as Automath connects so many aspects of logic,
mathematics and informatics, it may be worth-while to investigate
whether the teaching of mathematics could somehow profit from ideas that
emerged more or less naturally in the Automath enterprise. The idea of
Automath is to "explain things to a machine". Students are not machines
and should be approached in a different way. But as teachers we should
know that if we cannot explain a thing to a machine then we might have
difficulties in explaining it to students.

1.   In the Automath system the mathematical material is written in the
form of a complete book, line by line. A computer can check it line by
line, and once that has been done, the book can be considered as mathematically correct. The interpretation of such a book can be a complete
theory, containing all axioms, definitions, theorems and proofs.

2.   As a starting point we think of a book written entirely by human
beings. Later on we may think of leaving part of the writing to a
machine. That part might be simply the tedious routine work, but possibly also the more serious problem solving (i.e., "theorem proving", a
branch of artificial intelligence).

In order to be successful in the hard task of problem solving it might
be profitable temporarily to leave the format of the Automath languages.
In a way one might say that in this area generality and efficiency are
conflicting objectives. The Automath project made a choice here: it
never concentrated on automatic theorem proving, but just on checking.

3.   We should make a clear distinction between the Automath system and
Automath books. The system consists, roughly speaking, of language
rules and a computer program that checks whether any given book is
written according to those rules.

The system of Automath is mainly involved with the execution of substitution, with evaluation of types of expressions, and comparing such
types to one another. It is very essential that everything that is
said in a book, is said in a particular context: the context consists
of the typed variables that can be handled, but also of the list of
assumptions that can be used. The system keeps track of those contexts.

The Automath system does not contain any a priori ideas on what is
usually called logic and foundation of mathematics. Any logical system
(e.g., an intuitionistic one) can be introduced by the user in his own
book, and the same thing holds for the foundation of mathematics. In
particular, the user is not tied to the standard 20-th century set
theory (Zermelo-Fraenkel). And the user can choose whether or not to
admit things like the axiom of choice. From then on, the machine that
verifies the user's book will be able to do this according to the user's
own standards.

4.   In an Automath book, logic and mathematics are treated in exactly
the same way. New logical inference rules can be derived from old
ones, just like mathematical theorems are derived, and the new
inference rules can be applied as logical tools, in the same way as
mathematical theorems are applied.

5.   Writing in Automath can be tedious. All details of arguments have
to be presented most meticulously. At first sight this might be very
irritating. The questions are (i) whose fault this is, and (ii) what
can be done about it?

The questions are related. Part of the negative impression that the
length of an Automath book makes, is due to the fact that no attempt
was made to "do something about it" at the stage of the design of the
general system. This is based on the philosophy that generality comes
first, and that adaptability to special situations is a second concern.

The reason why Automath books become so long is that mathematicians
have more in their minds than they explain, and nevertheless we want to
handle all usual mathematical discourse. Perhaps we may say that part
of mathematical work is done subconsciously. Mathematicians have a
vast "experience" in mathematical situations, and such experience may
give a strong feeling for how all the little gaps can be filled. Possibly much of the experience is consulted subconsciously "on the spot".

Moreover, mathematical talking and writing are social activities. In
every area, people talk and write in a style they know they can get
away with. Some poor or incomplete forms of discourse are so widespread that it seems silly to bother about improvements; certainly it
is not a very rewarding task to try.

The answer to question (ii) is that very much can be done about it
indeed. But just as every user can write his own book under the
Automath system, he can implement his own attachments to the system.
This may involve special abbreviation facilities, but also automatized
text writing, producing packages of Automath lines by means of a single
command, in cases where there is a clear system behind such a package.

6.   Are computers essential for Automath? Not absolutely. The
computer sets the standard for what the notion "formalization" means.
If we cannot instruct a computer to verify mathematical discourse, we
have not properly formalized it yet. In the standard form, the author
of an Automath book has to write all the symbols one by one, and since
he knows that what he writes is correct, he would also be able to check
it by hand.

Nevertheless humans make mistakes. Automath books have been written
with a number of characters of the order of a million, all typed by
hand. It is hard to guarantee correctness of such a text without the
help of a modern computer.

7.   As the Automath system has no a priori knowledge of logic and set
theory, it can be used to write in a style that might be more natural
than what we see in other formalizations.

There is a wide-spread idea that propositional logic comes down to manipulating formulas in a boolean algebra, a kind of manipulation that is either carried out by handling formulas with the aid of lists of tautologies (in the same way as one used to do in trigonometry), or by a machine that checks all possibilities of zeros and ones as values for the boolean variables. A very much better formalization lies in the system of "natural deduction". This is very easy in Automath. The boolean bit-handling propositional logic can be done in Automath too, but it is much more clumsy than natural deduction.

A second option we get from the liberty of using Automath in the style we prefer, is to give up the 20-th century idea that "everything is a set". There is the magic Zermelo-Fraenkel universe in which every point is a set, and somehow all mathematical objects are to be coded as points in that universe. The particular coding is a matter of free choice: there is no natural way to code.

Zermelo-Fraenkel set theory is quite a heavy machinery to be taken as a basis for mathematics, and not many mathematicians actually know it. An alternative is to take "typed set theory", in which things are collected to sets only if they are of the same type: sets of numbers, sets of letters, sets of triangles, etc. It may take some trouble to make up one's mind about the question what basic rules for typed set theory should be taken as primitives, but if we just start talking the way we did mathematics before modern set theory emerged, we see that we need very little. Anyway, in Automath we have no trouble at all to talk mathematics in a sound old-fashioned way.

Yet, if someone still wants to talk in terms of Zermelo-Fraenkel universe, Automath is ready to take it.

8.   One of the advantages of Automath not being tied to any particular system for logic and set theory, is that we can think of formalizing entirely different things too, again in a natural style. As an example we may think of the algorithmic description of geometrical construc-tions like those with ruler and compass. Although it has not actually been produced, we may think of a single Automath book containing logic, mathematics and the description of ruler and compass constructions, with in particular the description and correctness proof (both due to Gauss) of the construction of the regular 17-gon. This description will be quite different from coding the construction as a point in the Zermelo-Fraenkel universe. We might even think of a robot equipped with ruler, compass, pencil and paper, who reads the details of the construction from the Automath book and carries them out in the way Gauss meant.

9.   Many parts of science are a patchwork consisting of pieces of theory, connected by rather vague intuitive ideas. Ever since the last part of the 19-th century it has been one of the ideas of the mathe-

matical community that mathematics should be integrated: all parts of
mathematics are to become sub-domains of one single big theory. The
patchwork picture still applies to most physical sciences, but also to
several parts of the mathematical sciences. One such part is informatics.

It seems to be a good idea to integrate informatics into mathematics,
at least in principle. And, as in the case of geometrical construc-
tions, Automath is a good candidate for describing this. It is possible
to write an Automath book containing: logic, mathematics, description of
syntax and semantics of a programming language, and particular programs
with proofs that the execution achieves the solution of particular
mathematical problems. One might even think of going further: description of the computer hardware with proof that it guarantees the
realization of the programming language semantics. Or directly, without
the intervention of a programming language, that a given piece of hardware produces a result with a given mathematical specification.

Needless to say, this kind of integrated theory will always contain a
number of primitives we have no proof for, but it will be absolutely
clear in the Automath book what these primitives are.

10. One thing people like in Automath, and other people strongly
dislike, is the way Automath treats proofs as if they were mathematical
objects. This is called "propositions as types". As the type of a
proof we have something that is immediately related to the proposition
established by that proof.

One should not be worried about this. Automath does not say that
proofs are objects, but just treats them syntactically in the same way
as objects are treated. This turns out to be very profitable: it
simplifies the system, as well as its language theory and the computer
verification of books. A third case where things are treated as
objects is the one of the geometrical constructions we mentioned in
section 8.

11. In standard mathematics, most identifiers are letters of various
kinds, possibly provided with indices, asterisks and the like. And
then there are the numerals, of course. We have learned from programming languages, however, to use arbitrary combinations of letters and
numerals as identifiers, (with restrictions like not to begin with a
numeral). We do the same thing in Automath, thus having the possibility to choose identifiers with a mnemonic value, like "Bessel",
"Theorem 137", "commutative". This certainly helps to keep books
readable.

In contrast to programming languages, the Automath system does not have
the numerals 0,1,...,9. One can introduce them as identifiers in a
book containing the elements of natural number theory, taking "0" and

"succ" (for "successor") as primitive, and defining 1:=succ(0),
2:=succ(1),..., 9:=succ(8), ten:=succ(9). After having introduced
addition and multiplication, we can define things like thirtyseven:=sum
(prod(3,ten),7), but the Automath system has no facilities to write
this as 37. This decimal notation might be added as an extra (it is
one of the possible "attachments" mentioned in section 5).

12. One of the basic aims of the Automath enterprise was to keep it
feasible. This has been achieved indeed: considerable portions of
mathematics of various kinds have been "translated" into Automath, and
the effort needed for this remained within reasonable limits. If we
start from a piece of mathematics that is sound and well understood, it
can be translated. It may always take some time to decide how to start,
but in the long run the translation is a matter of routine. As a rule
of thumb we may say there is a loss factor of the order of 10: it takes
about ten times as much space and ten times as much time as writing
mathematics the ordinary way. But it is not overimportant how big this
loss factor is (it would not be hard to reduce it by means of suitable
attachments, adapted to the nature of the subject matter). What really
matters is that it does not tend to infinity, which happens in many
other systems of formalizing mathematics. The main reason for the loss
factor being constant is that Automath has the same facilities for
using definitions (which are, essentially, abbreviations) as one has in
standard mathematics. The fact that the system of references is
superior to what we have in standard mathematics, makes it possible
that the loss factor even decreases in the long run when dealing with a
large book.

13. Another feature that makes Automath feasible is that we need not
always start at the beginning: we can start somewhere in the middle,
and if we need something that we have not defined, or have not proved,
we just take it as a primitive (primitive notion or axiom) and we go
on. We can leave it to later activity to replace all these primitives
by defined objects and proven theorems.

This kind of tactics was often (about 30 cases) applied at Eindhoven by
students (mathematics majors). It usually took the student not much
more than 100 hours work to learn about the system, to translate a
given piece of mathematics, to use the conversational facilities at a
computer terminal, and to finish with a completely verified Automath
book containing the result. In order to give an idea of the subjects
that had to be translated we mention a few: (i) The Weierstrass theorem
that says that the trigonometric polynomials lie dense in the space of
continuous periodic functions, (ii) The Banach-Steinhaus theorem, (iii)
The first elements of group theory.

14. Of the more extensive books that were written in Automath we
mention two. The first one is L.S. Jutting's complete translation of

E. Landau's <u>Grundlagen der Analysis</u>. In order to test the feasibility
of the system, the translator kept himself strictly to Landau's text,
rather than inventing some of the many possible shortcuts and improvements that would make the translation easier and shorter. The second
one we mention here was by J.T. Udding, who wrote a new text with about
the same results, much better suited to the Automath system, both in
its general outline and in its details. The gain over Landau's text,
in space as well as in time, was roughly 2.5.

15.   One of the ideas of the Automath enterprise was to get eventually
to a big mathematical encyclopaedia, a data bank, containing a vast
portion of mathematics in absolutely dependable form. This is a thing
that would take many hundreds of man years (thus far the Automath
project took something like 40). But the idea is feasible. Most of
the students mentioned in section 13 used the Landau translation (see
section 14) as a data bank, and that way they added to the bank.

16.   It is not the purpose of this paper to enter into details of the
Automath language, but the reader might want a general orientation.

There are several dialects of Automath in use, but here we only look
into basic things they have in common.

The expressions used are always lambda-typed lambda calculus expressions. This means that we have lambda expressions where every variable
has a type, and that type is again a lambda expression. In lambda
calculus the basic expression-forming devices are "abstraction" and
"application", but in Automath we have a further device, called
"instantiation". Instantiation is the operation that leads from an
n-ary prefix operator f to an expression f(E1,...,En), where E1,...,En
are expressions.

Having both "instantiation" and "application", Automath has two
different devices for expressing functionality, and both can be linked
to standard mathematical practice.

In Automath we write mathematics in the form of books, line after line.
There are three kinds of lines: (i) context lines, (ii) definitional
lines, and (iii) primitive lines.

A context line sets the context for the sequence of non-context lines
between that context line and the next one. A context is a sequence of
variables provided with types. We denote typing by a colon, and
describe a typical context of length 3:

$$x : A, \quad y : B(x), \quad z : C(x,y)$$

(here A, B(x), C(x,y) denote expressions; C(x,y) is an expression
containing the variables x and y).

A definitional line describes an abbreviation. It takes an expression E (of type F), and abbreviates that E by a new identifier, c, say. The line looks like

$$c := E : F.$$

A primitive line introduces some new identifier as a primitive notion, and attaches a type to it. That is, it is not defined, but declared to be available for further use. So it looks just like the definitional line above, but without the E. In order to stress that the defining expression E is omitted, we write a fixed symbol (like PN, or 'prim') in its place:

$$d := 'prim' : F.$$

These scanty remarks might give an idea about what the languages look like; for detailed description we refer to [1], and for an informal introduction into the use of the language also to [2].

We refer to [3] for a survey of the whole project, more extensive than the one given here. And also [4] will give a good idea about the project and the languages, but on top of that it is a report of all the experience obtained in the Landau translation mentioned in section 14.

REFERENCES

1   de Bruijn, N.G.   (1970) 'The mathematical language AUTOMATH, its usage and some of its extensions'. Symposium on automatic demonstration, Lecture Notes in Mathematics, Vol. 125, Springer Verlag.
2   de Bruijn, N.G.   (1973) AUTOMATH, a language for mathematics. (A series of lectures at the Sem. de math. sup., Un. de Montreal, June 1971.)   Lecture notes by B. Fawcett.   Les Presses de l'Un. de Montreal.
3   de Bruijn, N.G.   (1980) 'A survey of the project Automath'.   In: To H.B. Curry: Essays in combinatory logic, lambda calculus and formalism, ed. J.P. Seldin and J.R. Hindley, Academic Press.
4   van Benthem Jutting, L.S.   (1979) 'Checking Landau's Grundlagen in the Automath system', Mathematical Centre Tracts nr. 83, Amsterdam.

ON THE MATHEMATICAL BASIS OF COMPUTER SCIENCE.

Jacques STERN
Université de CAEN
CAEN, FRANCE

It is now clear to anybody that a working mathematician cannot ignore computers : as a consequence, it is commonly admitted that
students in mathematics, and especially those who intend to become teachers in the field, have to be exposed to some high-level language
(such as PASCAL). Nevertheless, this is far from being enough : the
question whether students in mathematics should be familiar with some
parts of the theoretical foundations of computer science cannot be
avoided because these topics are precisely the parts of computer science close to mathematics and seem to be necessary in order to establish
between both fields connections that go beyond the ability of using the
computing power of modern machines.

In France, following these line of ideas, an optional test
in computer science is now offered in the well-established "Concours
d'Agrégation de mathématiques". It is not my purpose to discuss the
role of the Agrégation in the French academic system : for those who
are not familiar with this system, let me simply say that the Agrégation
is a quite selective competitive exam which can be defined as a
"teaching Ph.D." and that many of the teachers for the age-group 17-22
have passed this exam. This is enough to understand why the fact that was
quoted above might be more important than appears at first glance : it
is likely that most of the requirements for the computer science test
of the Agrégation will become part of the standard curriculum leading
to graduation in mathematics. Although the author was not a member of
the group that defined these requirements, he agrees completely with
the choices of topics that have been made and is currently writing a
book covering these topics. The aim of the present contribution is precisely to give some general ideas that grew out during the first steps
of the process of writing this book. These ideas have to be considered
as my personal views on the subject although I owe a great debt to my
colleagues and especially to my co-author C. Puech, with whom I had
many inspiring discussions.

Before going into greater detail, I will make a last remark :
mastering some of the basic tools of computer science will not turn a
mathematician into a computer scientist. Instead, it should help to develop a different frame of mind, suitable to understand the specific
features of computer science. This is most important for a mathematician

because, as is shown in other contributions of this book, these specific features will necessarily react both on the teaching and on the practice of mathematics themselves.

1   AROUND THE NOTION OF COMPUTATION

Computation theory is considered by many people as a very dull subject ; nevertheless, it is the first burden of the theory to provide a suitable criterion for drawing a limit between what is computable (or effective) and what is not. A simple way would be to use the word computable for everything that can be processed on a real computer. Although this point of view is not completely meaningless, it remains rather vague and cannot be considered as a genuine mathematical notion because of its lack of precision. Furthermore, this point of view is not even historically correct ; a lot of outstanding work connected with the subject of computation theory was published before the first modern computer was built : for example, one can quote the work of Turing [1936], Kleene [1936], Post [1936] on computation itself but also the work of Mc Culloch and Pitts on the modelling of neuron nets, from which grew out the theory of automata.

It is precisely the theory of automata that we propose to choose as a starting point. Many reasons can be put forward in order to justify such a choice. The theory is simple and firmly established and provides various exercises in programming : for example, one can simulate an automaton in a high-level language like PASCAL. Next, it provides several opportunities to write simple algorithms and even to speak rather informally of their complexity (e.g. by comparing various algorithms for computing the minimal automaton). Also, the concept of non-determinism can be introduced naturally and in a simple setting. Finally, automaton theory has several applications : to text editors and to lexical analysis in particular ; this is not a minor argument.

Nevertheless, one quickly comes to the conclusion that automata do not provide a satisfactory model for real machines. This conclusion can be reached by writing down simple languages that are not accepted by a finite automaton but also through the simple and more convincing observation that a central feature of computers is completely wiped out : the ability to store data in a memory. We are back to our original problem of defining the notion of "computable" and it is reasonable, at this point, to require that this notion should be isolated by various different techniques that come out to be equivalent : this will show that a mathematical invariant has been found and this will make Church's thesis (that identifies "computable" with "recursive") highly plausible.

Four different approaches can be taken.

1) Adding a memory device to a finite automaton.   This yields the definition of a Turing machine.

2) <u>Directly modelling actual computers</u>. This can be done through the notion of random access machine (cf. Cook and Reckhow [1973]) operated by a very simple language similar to machine language.

3) <u>Defining a simple class of programs</u>. For example one can define a restricted version of PASCAL which uses only the integer type and the control structures IF...THEN...ELSE and WHILE...DO.

4) <u>Defining the class of (partial) recursive functions</u>. This is a good opportunity to discuss functional languages: recursive definitions can be handled by using some constructs similar to those appearing in LISP.

The proof that all these definitions are actually equivalent is a source of very interesting observations. For example, the fact that the restricted PASCAL can compute all recursive functions proves the well-known fact that the GOTO statement can be dispensed with. It may be worthwhile to note that replacing WHILE...DO by FOR only allows the computation of primitive recursive functions. Also, the simulation of a random access machine by a Turing machine is a good exercise that shows how to handle a sequential memory. A further exercise would be to use non-erasing Turing machines: this could be motivated by comparing these Turing machines with the write-only memories that are offered by the recent technology of optical disks.

Once the notion of a computable function has been given a precise definition, it becomes possible to discuss decidability issues: the construction of a universal machine does not require much more effort and the "halting problem" can be properly stated and studied. Then, one can open a discussion on whether or not the dichotomy decidable/undecidable is of practical importance. This is a way to introduce complexity theory. Going back to the simulation of one machine by another, one can check that polynomial time has a stable character. This allows the definition of the class P which is a reasonable candidate for modelling a class of problems sometimes called "feasible" or "tractable".

2 <u>AROUND THE NOTION OF ALGORITHM</u>

Now that we are equipped with a theoretical notion of complexity, it is necessary to use it in more concrete situations. This can be done through a review of various algorithms. This review is, by no means, an exercise in programming style even if correct programs have to be written at some point. The emphasis should be on the design and evaluation of these algorithms, which are very closely connected. Of course, the rules of the game should be clearly stated and discussed, especially the choice of a discrete model and the main notions of complexity that are in use: worst-case analysis and average-case analysis. The choice of the method depends on the underlying probabilistic model: average analysis is relevant when the probability of "ill-behaved" cases is small. In any case, the analysis is combinatorial in character and exact formulae are not easily handled even if they can be written down: some specific tools are needed like statistics of permutations and distributions and the use of generating series (cf. Knuth [1973]). Generally, such techniques (e.g. the use of singular points of the generating series) only allow an asymptotic analysis and one may ask if this kind of information has any practical meaning ; after all, the size of the data are bounded by the computing environment ! It turns out that the asymptotic analysis is actually relevant : when a given algorithm runs in time $O(n \log n)$ for example, it is usually true that the constant implicit in the O notation is rather small and that the asymptotic inequality is reached rather quickly.

Together with algorithms, the specific data structures used in computer science should be discussed : stacks, files, trees, graphs... It should be stressed that this point of view is quite different from the one that was taken in §1 ; in computation theory, we made various simulations involving basic manipulations on data-structures and we claimed that these manipulations were not costly. In practical issues, the choice of good data-structures may actually save a significant part of the running-time and has to be made carefully.

We now briefly comment on some specific choices that can be made ; of course, what follows is not an exhaustive list.

1) <u>Sorting</u>. There is a very large number of sorting algorithms ; a few can be studied and compared e.g. insertion sort, heapsort, quicksort. This is enough to make some quite interesting observations. First of all, the actual evaluation of algorithms does not really use the original definition of complexity that was given for random access machines. In practical situations, the size of the integers is bounded and the complexity is roughly the number of machine instructions performed during execution. Even further simplifications are made : for example, some instructions are ignored (especially those concerning data manipulations) ; thus, in sorting, one is left with the number of comparisons as a meaningful measure of complexity. Also, sorting algorithms give an opportunity to establish the difference between worst-case complexity and average-case complexity. The efficiency of quicksort is also a nice illustration of a basic but very powerful principle of computer science : the <u>divide-and-conquer</u> method.

2) <u>Searching</u>. Sequential search, binary search, hashing can be discussed but the emphasis here should be on specific data structures that can be considered : binary trees, A.V.L. trees or 2-3 trees. The use of these structures should make the student familiar with the important concept of <u>balancing</u>, which is another illustration of the divide-and-conquer method.

3) <u>Pattern matching</u>. The Knuth-Morris-Pratt algorithm can be described. This is mainly included because of the connection with automaton theory, which was studied at length in the first part.

4) <u>Graph algorithms</u>. This is a very interesting part because it clearly illustrates the interplay between mathematics and computer science and shows how the way to look at mathematical objects can be

affected. Graph theory can be viewed as a part of mathematics of its own ; definitions can be given and nice theorems can be proved, for example on the existence of spanning trees. On the other hand, the actual computation of a minimum-cost spanning tree requires some extra work. This new point of view, in its turn, raises deep questions of a mathematical nature connected with the complexity of the algorithms.

It should also be added that basic algorithms that compute the shortest path or the transitive closure of a graph are quite efficient and quite useful. On the other hand it should be mentioned that, for many problems (such as the hamiltonian path problem) no polynomial-time algorithm is known ; this is a good introduction to the subject of non-polynomial algorithms.

The class of NP and NP-complete problems can be given a precise definition through nondeterministic Turing machines or nondeterministic random access machines. The main task here is to establish Cook's theorem (cf. Cook [1971]). Once this is done, a sample of NP-complete problems can be given (cf. Garey, Johnson [1978]), such as :

> The satisfiability problem
> The travelling salesman problem
> The hamiltonian path problem
> The clique problem
> The knapsack problem.

Finally, some indications can be given on how to handle NP-complete problems : for example, one can show how minimum-cost spanning trees can be used to find a "good" solution to an instance of the travelling salesman problem.

3   AROUND LOGIC: SYNTAX AND SEMANTICS

The theory of context-free languages offers a first example of a syntactical approach, through context-free grammars. Of course, the connection with pushdown automata should be made precise. Once this is done, applications of context-free grammars can be given e.g. to syntactical analysis.

The study of derivation trees for context-free grammars is also a way to get the student ready to understand the basic tools of logic such as the rules of inference. Surprisingly, those tools are sometimes a cause of panic for mathematicans ! Of course, one should choose to develop logic (and especially the completeness theorem) in a very constructive way, which is suitable for computer science. The role of Skolem functions in restricting one's attention to $\forall \exists$ formulae should be explained ; Herbrand's theorem and Robinson's resolution algorithm should play a central role (cf. Robinson [1965]). Of course, undecidability should be clearly discussed : the Herbrand procedure does not necessarily come to a stop. Nevertheless, the usefulness of resolution can be stressed by a quick overview of the PROLOG language.

Once those basic topics from logic have been covered one can briefly discuss some more advanced matters such as

1) Semantics, especially the semantics of recursive procedures and the fix-point approach of programs.

2) Program verification through Hoare's logic (cf. Hoare [1969]).

CONCLUSION

In this short paper, we have tried to describe what we consider as the mathematical basis of computer science, to show how the chosen topics can be organized and to motivate the choices that have been made. Clearly, those contents will change very quickly, following the further development of computer science. Maybe some mathematical tools for the study of VLSI or of relational databases will have to be added soon. In any case we feel that mathematical tools for computer science should become a part of the advanced curriculum in mathematics.

REFERENCES

Aho A.V., J.E. Hopcroft and J.D. Ullman [1974]. The Design and Analysis of Computer Algorithms, Addison Wesley, Reading, Mass.

Aho A.V., J.E. Hopcroft and J.D. Ullman [1983]. Data Structures and Algorithms, Addison Wesley, Reading, Mass.

Aho A.V. and J.D. Ullman [1977]. Principles of Compiler Design, Addison Wesley, Reading, Mass.

Cook S.A. [1971]. 'The complexity of theorem proving procedures'. Proc. Third Annual ACM Symposium on the Theory of Computing, 29-33.

Cook S.A. and R.A. Reckhow [1973]. 'Time bounded random access machines', J. Computer and Systems Science 7, 354-375.

Garey M.R. and D.S. Johnson [1978]. Computers and Intractability, a Guide to the Theory of NP-Completeness, H. Freeman, San Francisco.

Hoare C.A., [1969]. 'An Axiomatic Basis of Computer Programming', C. ACM 12, 576-580.

Kleene S.C. [1939]. 'General recursive functions of natural numbers', Mathematische Annalen 112, 727-742.

Knuth D.E. [1973]. The Art of Computer Programming: Vol. III: Sorting and Searching, Addison Wesley, Reading, Mass.

Manna Z. [1974]. Mathematical Theory of Computation, McGraw Hill, New York.

McCulloch W.S. and W. Pitts [1943]. 'A logical calculus of the ideas immanent in nervous activity', Bull. Math. Biophysics 5, 115-113.

Post E. [1936]. 'Finite combinatory processes-formulation I', J. Symbolic Logic 1, 103-105.

Robinson J.A. [1965]. 'A Machine-oriented Logic Based on the
        Resolution Principle', J. ACM 12, 23-41.
Rogers H. Jr. [1967]. Theory of Recursive Functions and Effective
        Computability, McGraw Hill, New York.
Sedgewick R. [1983]. Algorithms, Addison Wesley, Reading, Mass.
Wirth N. [1976]. Algorithms + Data Structures = Programs, Prentice
        Hall, Englewood Cliffs, N.J.

MATHEMATICS OF COMPUTER ALGEBRA SYSTEMS

J. H. Davenport, School of Mathematics, University of Bath, Bath, BA2 7AY, England

INTRODUCTION

ICMI [1984] poses the question "What is the mathematics underlying symbolic mathematical systems"? The aim of this paper is to give some answers to this question, and also to address the following question that ICMI does not directly answer: "How does this mathematics relate to current curricula", which could be re-phrased as "What aspects of current curricula are rendered obsolete, or drastically changed by symbolic mathematical systems". It should be emphasised that this paper does *not* address the question "How should algebra systems be used to teach existing mathematics in the same way", though that is a very important question.

ELEMENTARY CALCULATIONS

Symbolic mathematical systems are capable of a variety of essentially trivial calculations. An obvious example is the multiplication of polynomials. The algorithm for doing this is taught at school, and there is little doubt that any competent student knows how to multiply polynomials. He may make a mistake while doing so, but that would be an accident, and he would recognise the mistake if it were pointed out to him. This does not mean that the student could actually do the calculations. They may well be too long for him, either in terms of time or in terms of the probability of there being an error.

Either the student or the experienced mathematician may wish to use a computer algebra system to multiply polynomials. Andrews [1979] used one to multiply four polynomials together to verify a 752-term identity. The student may wish to use them for easier examples. Indeed, if the student is to use these systems at all, he will start via calculations that are elementary in principle.

A remark that is important for joint Mathematics and Computing degrees, though not necessarily for purely Mathematics ones, is that these algorithms really are easy to program in a suitable language. PASCAL, and *a fortiori* FORTRAN and BASIC, are not suitable because of the problems of storage management, but in LISP, for example, a polynomial multiplication and addition package can be written in under 50 lines, and division is not much more complicated. Similarly, Billard [1981] reports that addition and multiplication routines take 2.8 kilobytes in APL. The author has used these problems to good effect in an introductory LISP programming course for applied mathematicians, and many colleagues have done much the same.

Polynomial greatest common divisors are a more interesting case. The obvious algorithm is Euclid's, but this has several draw-backs. The direct implementation forces us to work with polynomials over the rational numbers, and this gives us many calculations with rational numbers, and much effort in computing the g.c.d. of integers in order to simplify these rationals.   Hence the obvious variant, which is to clear denominators at each step, or, more succinctly, to cross-multiply rather than create fractions.   This would seem an eminently sensible algorithm, and one hallowed by tradition.   As was pointed out by Brown [1971], though, computation of the greatest common divisor of $x^8+x^6-3x^4-3x^3+8x^2+2x-5$ and $3x^6+5x^4-4x^2-9x+21$ gives rise to numbers with 35 decimal digits.   This appears ridiculous, since the Landau-Mignotte inequality [Mignotte, 1974] assures us that no coefficient in the g.c.d. can possibly be greater than 480.   When it comes to polynomials in several variables, it is equally possible for the degrees in the "non-main" variables to increase absurdly. In an effort to combat this *intermediate expression swell*, there has been much investigation of alternative algorithms.

ADVANCED METHODS

These methods all have the same general outline, expressed by the following diagram, which we give for the case of greatest common divisors in $Z[x]$.

$$
\begin{array}{ccc}
Z[x] & \xrightarrow{\text{g.c.d.}} & Z[x] \\
\phi \downarrow & & \downarrow \phi^{-1} \\
D & \xrightarrow{\text{g.c.d.}} & D
\end{array}
$$

In essence since the problem is too difficult in $Z[x]$, the problem is mapped to a domain D, solved there, and then the mapping is inverted to give a solution in $Z[x]$.   In all cases, D is chosen so as to be "easier" than $Z[x]$. Historically, the first choice was $Z_N[x]$ where N is the product of distinct primes. In fact, the calculation in D is performed by computing the greatest common divisor modulo each prime separately, and then combining the results via the Chinese Remainder Theorem. ϕ is obvious, and, if N is large enough, then the answer in D should also be the answer in $Z[x]$. This is the *modular* method of Brown [1971].

The next method, the p-adic method, also takes D to be $Z_N[x]$, but now N is p^n for some prime p.   Here the g.c.d. is first computed modulo p, and then lifted to p^2, p^3, ... (or p^2, p^4, p^8, ...) by Hensel's Lemma.   This method was pioneered by Yun [1974], and has since undergone many variations [Wang, 1980; Zippel, 1981].

A more recent method, the Z-adic method [Char *et al.*, 1984; Davenport & Padget, 1985], takes a somewhat different approach.   Here D is taken to be Z.   The computation in D is merely that of an integer g.c.d., and ϕ is the operation of replacing x by some integer n. ϕ^{-1} is less obvious: we write a number N in the base n, but with digits between $-n/2$ and $n/2$, and then regard each digit as the coefficient of the corresponding power of x.

These methods all have the feature that the diagram need not commute, i.e. that $\phi(\text{g.c.d.}(p,q))$ need not be equal to $\text{g.c.d.}(\phi(p),\phi(q))$, e.g. if $p = x-1$ and $q = x+2$, then p and q are relatively prime over the integers, but are equal, and hence have themselves as a common factor, modulo 3.   The techniques for detecting such *bad evaluations* vary from method to method, and are not of great interest here: the reader is referred to the detailed references.   It is worth remarking that these methods are not limited to univariate g.c.d. calculations: they all have analogues for multivariate problems.

Such uses of homomorphisms and backward reasoning from homomorphic images are very common in research mathematics, but are rarely seen at the undergraduate level.   Indeed, when teaching this material to well-trained mathematics students, the author has often had the comment "Oh, so *that's* what homomorphisms are for".   This material can be taught to students with some background in abstract algebra, or the abstract algebra can be taught with these methods as motivation, as is done by Lipson [1981].

FACTORISATION

When it comes to the factorisation of polynomials, there is no "easy" method, and guess-work is the usual method of attack taught.   This can be systematised in "Kronecker's method" (actually due to Newton and von Schubert), but is hideously expensive in general.   Hence we need one of the "advanced" methods mentioned above.   Since factoring integers is also very expensive, we can rule out the Z-adic method.   We first remark that we can restrict ourselves to the case of square-free polynomials, since square-free decompositions can be calculated by means of g.c.d.s.   The method we shall describe is essentially due to Zassenhaus [1969], though many authors have worked on its implementation and refinement (see Kaltofen [1982] for a survey).

If a polynomial f factors over the integers, then it certainly factors modulo every prime p (unless the leading coefficient vanishes, in which case one of the factors might vanish).   The converse of this is that if f is irreducible modulo some p, then it is irreducible over the integers. Similarly, we can conclude that it is irreducible if the factorisation modulo two different primes are incompatible, e.g. $(x^3+\ldots)(x^3+\ldots)$ modulo one prime and $(x^4+\ldots)(x^2+\ldots)$ modulo another.   Musser [1978] suggests trying five different primes in an attempt to demonstrate irreducibility.

Unfortunately, such methods assume that we can factorise polynomials modulo p, and this is not obvious either.   The great breakthrough was made by Berlekamp [1967], who produced an efficient algorithm for this, whose running time is $O(n^3+prn^2)$ for factoring a polynomial of degree n modulo p, where r is the number of factors.   The algorithm involves the matrix Q, whose rows are the coefficients of y^i modulo the polynomial to be factored, where y is x^p, but it is too complicated to describe here.

It is unfortunately possible for an irreducible polynomial to factor consistently modulo every prime p: e.g. x^4+1 factors as the product of two quadratics (which may or may not be irreducible) modulo every prime, and Swinnerton-Dyer [see Berlekamp, 1967] has generalised this phenomenon. Hence, unlike the previous section, we cannot just rely on good evaluations, but we have also to make use of bad evaluations.   Assuming that the polynomial remains square-free, we can use Hensel's Lemma to produce a factorisation modulo p^n for any power we choose.   We can choose a power more than twice as large as any possible coefficient in a factor, by the Landau-Mignotte inequality [Mignotte, 1974].   Hence if the factors modulo p^n correspond to genuine factors over the integers, they are equal to those factors.   If not, then we must try all possible combinations in order to look for factors which split further modulo p.   This combinatorial search means that the worst case of this algorithm is exponential in the degree of the input.   More recently, other algorithms [Lenstra *et al.*, 1982] have been devised that avoid this step, though in practical cases they may actually be more expensive.

It is clear that this is a complex algorithm, relying on a fair amount of relatively advanced algebra.   In this respect, it goes beyond the modular g.c.d. algorithm of the previous section.   One can explain this algorithm in all its detail: indeed the author has done so easily and successfully to people with one year's post-graduate training in Number Theory.   But this is an exception, and nearly all people who use such an algorithm have little understanding of how it works, nor need they have. Factorisation is an idea which is far easier to grasp than to compute, and why not leave the computation to the computers and the specialists?

INTEGRATION

Factorisation is nice, but generally speaking (see, however, Coppersmith & Davenport [1985] for an example where factorisation is the key technique) it is a bonus rather than a necessity.   Integration, on the other hand, has major uses throughout applied mathematics.   Despite this, and despite many well-established results on the existence or non-existence of closed forms, the methods of integration taught in schools and universities are largely heuristic: look for a substitution, or sequence of substitutions, that makes the integral look like one that you know, or can find in an integral table. Errors are quite possible on these substitutions, and integral tables are not perfect: Klerer & Grossman [1968] quote typical error rates in excess of 10%, with a peak over 25%.

Much of the theory of integration is actually relatively straight-forward: we need to start with some definitions and a theorem stated *ex cathedra*.   A function is said to *elementary* if it is built up from constants and the variable of integration by means of the arithmetic operations, the taking of exponentials and logarithms, and the solution of algebraic equations.   Hence log x, $\sqrt{(1+\exp(x))}$, sin x (which is $(e^{ix}-e^{-ix})/2i$) and many other functions are elementary.   The *field of definition* of such a function is the smallest field closed under differentiation that contains this function and the building blocks (exponentials, logarithms and algebraic solutions) used to define it.

Theorem [Liouville, 1835].   Let f be an elementary function whose field of definition is F.   Then, if f has an elementary integral, i.e. an elementary function g such that $g' = f$, then g can be written as

$$g = v_0 + \sum_{i=1}^{n} c_i \log v_i \ , \quad \text{so that } f = v_0' + \sum_{i=1}^{n} \frac{c_i v_i'}{v_i} \ , \tag{1}$$

where v_0 belongs to F, the c_i are constants satisfying algebraic equations over F and the v_i belong to $F(c_1, \ldots, c_n)$.

The proof of this theorem is not terribly deep, though it requires a fair amount of work to make the various notations involved precise, and deduce formally various "intuitive" lemmas. The proof is essentially a formalisation of the remark "differentiation does not remove exponentials or algebraic functions, and can only remove logarithms if their coefficients are constant".

Given this theorem, it is an easy matter to prove, for example, that $\exp(-x^2)$ has no elementary integral. If we write θ for $\exp(-x^2)$, then the field of definition can be taken as $Q(x, \theta)$. Hence the problem is to express θ in form (1). Now, if v_0 is P/Q, where P and Q belong to $Q(x)[\theta]$, then the denominator of v_0' will contain either θ (from a factor of θ in Q) or repeated factors (since the derivative of $1/r$ is $-r'/r^2$). But the denominator of v_i'/v_i will contain only non-repeated factors, and no factor of θ (since $\log \theta$ is merely $-x^2$). Hence there can be no cancellation between the denominators of v_0' and the other terms. But the left-hand side, θ, has no denominator, and hence we deduce that there are no new logarithms, and that v_0 has no denominator. This leaves us with the equation

$$\theta = \left[\sum_i a_i \theta^i \right]' = \sum_i a_i' \theta^i - 2x \sum_i i a_i \theta^i$$

where the a_i are elements of $Q(x)$. Equating coefficients in this shows us that $1 = a_1' - 2xa_1$.

a_1 belongs to $Q(x)$: write it as p/q, where p and q are both polynomials. Let n be the greatest multiplicity of any root of q (if any). Then the denominator of a_1' has at least one root of multiplicity $n+1$, while the denominator of $2xa_1$ has roots of multiplicity at most n. Hence the denominator of the right-hand side has a root of multiplicity $n+1$, which is impossible, since the left-hand side is 1. Hence there are no roots of q, and q is a constant, i.e. a_1 is a polynomial. Let m be the degree of this polynomial. Then $m-1$ is the degree of a_1', while $m+1$ is the degree of $2xa_1$. Hence $m+1$ is the degree of a_1-2xa_1: again a contradiction. This, in fact, contradicts the initial assumption that θ had an elementary integral.

The previous two paragraphs have shown, via an *ad hoc* argument, that $\exp(-x^2)$ has no elementary integral. It is a remarkable fact that this process can be rendered formal and algorithmic, and indeed that the following result is true.

Theorem [Risch, 1969]. There is an algorithm which, given an elementary function lying in a purely transcendental extension of $Q(x)$, will decide whether or not this function has an elementary integral.

The algorithm referred to is essentially a systematisation of the method we have seen, by equating coefficients one variable at a time. This does not, of course, mean that the proof is trivial, and indeed there is a substantial amount of subtlety involved in showing when cancellation can not occur, and in bounding it when it can occur.

Why should these methods not be taught in schools in place of the present "guess a substitution, or integrate something by parts" techniques? They may place a slightly greater demand on the students ability to absorb abstract theorems, but they certainly place far less strain on the memory than a large number of substitutions and known results.

There is also an alternative technique, which is essentially a systematisation of integration by parts, due to Risch & Norman [Davenport, 1982]. This is even more straight-forward to explain, since it does not have the recursive step of the other, though its theoretical foundations are less useful, since in general we cannot prove that not finding an integral implies the non-existence of an integral. The method argues that the integral of $\exp(-x^2)$ must be $\sum a_{ij}x^i\exp(-x^2)^j$. The derivative of such a sum (which must be equal to the integrand) is clearly $\sum a_{ij}ix^{i-1}\exp(-x^2)^j + \sum -2ja_{ij}x^{i+1}\exp(-x^2)^j$. The leading term is the second one, which contains x^{i+1}. This clearly cannot match the integrand, which is $x^0\exp(-x^2)^1$.

In general, to integrate a rational function of several functions (which we shall call the *kernels*), such as x, $\exp(-x^2)$ or $\log(x)$, write the function as p/q, where p and q are polynomials in these kernels. We now expect that the integral will be $r/s + \sum c_i\log(v_i)$, where the c_i are unknown constants, r is an unknown polynomial in the kernels, s is a known polynomial in the kernels (the product of all the factors of q, with their multiplicities decreased by 1) and the v_i are known polynomials (the irreducible factors of q, with perhaps some extra ones thrown in). Hence $p/q = (r/s)' + \sum c_i v_i'/v_i$. We can clear denominators in this equation, and we are left with a system of linear equations to solve for the c_i and the coefficients of r. Furthermore these equations are generally banded, and can be solved quite simply. For complicated integrals, the manipulation mentioned above is lengthy, but it can certainly be explained quite simply, and, with a computer to do the dirty work, students can see how the integral is deduced.

PHYSICAL APPLICATIONS

Much of Physics and Engineering is based on mathematical models and structures. Unfortunately, these models may be too complicated to tackle. This happens at every level, from the school teacher who pretends that the mass of the rocket is fixed, through the university lecturer who does relativity in one space dimension, to the research astronomer who treats cylindrical stars because spherical ones are too difficult to model.

Sometimes the difficulties are genuine, and often an artificially simple case can, because of its simplicity, convey more insight. Sometimes the difficulties are purely manipulative, and the use of an algebra system can make it possible to consider a more realistic case than that which is possible by hand.   For example, it is possible to consider orthogonal polynomials in various co-ordinate systems, rather than being limited to the two or three which students know about, and, in practice, to the one or two which they can actually manipulate.

Just as it is possible for a lecturer to perform a calculation on a calculator which the students can check, so it is possible to perform a calculation on a computer algebra system, though the practice requires some facility to broadcast the computer screen throughout the lecture hall. In fact, running a script through an algebra system, is, if the script is well-commented and available to the students, a better pedagogic tool than covering the blackboards in algebra and leaving the students to distinguish the insight from the manipulation.

CONCLUSION

In some cases, computer algebra can illuminate a piece of standard mathematics.   For example: "Why did the computer know so quickly that these polynomials were relatively prime" might have the answer "Because they are relatively prime modulo 3".   This sort of discussion motivates homomorphisms, and, at a more advanced level, the subtle algebraic-geometric concept of *good reduction*.

In other cases, computer algebra can provide a better way of tackling a standard problem.   Many examples of integration are better tackled by explaining (or at least quoting) the theory than by producing a substitution "out of a hat" for solving the particular problem, which gives no general insight.

In a third class of cases, such as factorisation, there is a general theory available in computer algebra systems which is essentially beyond the reach of most students who will need to use it.   This does not matter, since the computer algebra implementation of this theory can be treated as a black box.   As an analogy, how many of the students who use a calculator to determine a square root could compute it for themselves, although they know that it satisfies $(\sqrt{x})^2 = x$?   It would clearly be better if the students could all compute it, but the constraints of a finite timetable and finite interest among the students force school teachers to regard the calculator as largely a black box, and computer algebra systems can always be regarded as such if the situation demands it.   This attitude will render it unncessary to teach, and to practise, students in "finding simple roots", since the computer will be able to do that for them.

This is not to say that this matter is trivial, or that all teachers will find that computer algebra solves their most pressing problems.   Indeed, it is reasonable to predict that computer algebra will cause high school and university teachers at least as much trouble as calculators caused their colleagues teaching younger pupils.   Since computer algebra systems can tackle a wider range of more sophisticated problems, they will probably cause more havoc.   But this cannot be avoided.   For better or for worse, if the technology is there, the pupils will use it.   The author has already seen algebra systems used by undergraduates doing numerical analysis, and the availability of algebra systems on the IBM PC and its relations, and soon on the Mackintosh, means that the students will have access to the technology. Hence the availability of computer algebra systems is, like the calculator, both an opportunity and a threat to the teacher. Opportunity because he can teach much more, threat because so much that he teaches is, or seems to be, redundant.

In sum, the answers to the questions posed above seem to be as follows. *What are the mathematics underlying symbolic mathematical systems?* Modular, p-adic and Z-adic methods; the sophisticated use of various bounds, and a specialised theory of integration. *How does this mathematics relate to current curricula?* Some of it develops ideas, like homomorphism, that are in current curricula, some of it, like the theory of integration, could replace much of the material taught, and some of it, like factorisation, is basically too advanced.

REFERENCES

Andrews, G.E., *Plane Partitions (III): The Weak MacDonald Conjecture.* Invent. Math. 53(1979) pp. 193-225.

Berlekamp, E.R., *Factoring Polynomials over Finite Fields.* Bell System Tech. J. 46(1967) pp. 1853-1859.

Billard, B., *Polynomial Manipulation with APL.* Comm. ACM 24(1981) pp. 457-465, 25(1982) p. 213.

Brown, W.S., *On Euclid's Algorithm and the Computation of Polynomial Greatest Common Divisors.* J. ACM 18(1971) pp. 478-504.

Char,B.W., Geddes, K.O. & Gonnet, G.H., *GCDHEU: Heuristic Polynomial GCD Algorithm Based on Integer GCD Computation.* Proc. EUROSAM 84 (Springer Lecture Notes in Computer Science 174) pp. 285-296.

Coppersmith, D. & Davenport, J.H., *An Application of Factoring.* To appear in J. Symbolic Computation 1(1985) no. 2.

Davenport, J.H., *On The Parallel Risch Algorithm (I).* Proc. EUROCAM 82 (Springer Lecture Notes in Computer Science 144) pp. 144-157.

Davenport, J.H. & Padget, J.A., *HEUGCD: How Elementary Upperbounds Generate Cheaper Data.* To appear in Proc. EUROCAL 85 (Springer Lecture Notes in Computer Science).

International Commission on Mathematical Instruction, *The Influence of Computers and Informatics on Mathematics and Its Teaching.* L'Enseignement Mathematique 30(1984) pp. 159-172.

Kaltofen,E., *Factorization of Polynomials.* Symbolic and Algebraic Computation (Computing Supplementum 4) (ed. B. Buchberger, G.E. Collins & R. Loos), Springer-Verlag, Wien-New York, 1982, pp. 95-113.

Klerer, M. & Grossman, F., *Error Rates in Tables of Indefinite Integrals.* Industrial Math. 18(1968) pp. 31-62.

Lenstra, A.K., Lenstra, H.W., Jr. & Lovasz,L., *Factoring Polynomials with Rational Coefficients.* Math. Ann. 261(1982) pp. 515-534.

Liouville, J., *Memoire sur l'integration d'une classe de functions transcendantes.* Crelle's J. 13(1835) pp. 93-118.

Lipson, J.D., *Elements of Algebra and Algebraic Computing.* Addison-Wesley, 1981.

Mignotte, M., *An Inequality about Factors of Polynomials.*   Math. Comp. **28**(1974) pp. 1135–1157.

Musser, D.R., *On the Efficiency of a Polynomial Irreducibility Test.* J. ACM **25**(1978) pp. 271–282.

Risch, R.H., *The Problem of Integration in Finite Terms.* Trans. A.M.S. **139**(1969) pp. 167–189.

Wang, P.S., *The EEZ-GCD Algorithm.*   SIGSAM Bulletin **14**(1980) pp. 50–60.

Yun, D.Y.Y., *The Hensel Lemma in Algebraic Manipulation.* Ph.D. Thesis and Project MAC TR–138, M.I.T., 1974. Garland Publ. Corp., New York, 1980.

Zassenhaus, H., *On Hensel Factorization I.*   J. Number Theory **1**(1969) pp. 291–311.

Zippel, R.E., *Newton's Iteration and the Sparse Hensel Algorithm.* Proc. SYMSAC 81, ACM, New York, pp. 68–72.

MATHEMATICAL EDUCATION IN THE COMPUTER AGE

Haruo Murakami
Faculty of Engineering and The Graduate School of
Science and Technology,

Masato Hata
The Graduate School of Science and Technology,
Kobe University, Japan

1 INTRODUCTION

The progress of computers has been remarkable. Today, along with the conventional fast numerical computation, the extensive use of computers for non-numeric operations has begun in a variety of fields. Such non-numeric operations in mathematics include computer algebra allowing symbolic differentiation, symbolic integration, factorization and expansion, etc.

The astonishing speed of technological innovation will soon make possible a cheap computer algebra system as small as present-day electronic calculators or hand held computers, but with mathematical capabilities as powerful as those of average first-year college students at least as far as the above symbolic mathematical operations are concerned. Most of the powerful systems such as MACSYMA or REDUCE operate in minicomputers or larger ones now. But technological innovation is realizing much smaller computers capable of doing the same functions. In fact, REDUCE has recently been implemented on a personal computer using MC68000.

The emergence of such powerful but small computer algebra systems will inevitably influence mathematical education. Therefore, it is now very important to consider the following questions:

What influence will computers with the above capabilities have on mathematical education?

How will mathematical education be changed by computers?
Or, how will it have to be changed?

2 THE INFLUENCE OF COMPUTERS ON MATHEMATICAL EDUCATION

The changes which computers will bring in mathematical education can be divided into:

i) changes in the methodology of mathematical education,
ii) changes in the topics taught in mathematical education.

First, we consider the changes in the methodology of mathematical education. Many CAI (Computer Assisted Instruction) systems have already been tested. The results show that typical CAI systems using

the drill and practice mode are effective in improving students'
ability to do formal calculations and in helping students to understand
a new concept or topic by the use of graphical images.

This type of application will be widely used in mathematical education.
This being the case, the methodology of mathematical education will
have to be changed. Namely, the conventional methodology whereby a
teacher teaches everything by him/herself will be replaced by a new one
in which the teacher can selectively use computers for a particular
topic or situations in which computers are very effective. It will,
therefore, be necessary to establish a new methodology in order that
computers can be used most effectively.

It is noted that most of the CAI systems which have been developed are
designed for the acquisition of mathematical knowledge and computational
skills. This objective is one of the two distinct objectives of
mathematical education. The other is the acquisition of the capacity
for mathematical (logical) thought. However, few CAI systems have been
developed yet for the second objective. This is due to the fact that
the methodology to develop the second objective is not explicitly
established. Therefore, it is concluded that we cannot let computers
replace a greater part of what the human teacher has been teaching until
the methodology for the second objective is explicitly established and
a CAI system based on it is developed.

Next, we consider the changes in the topics taught in mathematical
education. What to teach is determined by the demands of society. As
computers increase in importance, the demands of society change. Thus,
there may be an increase in: 1) the demand for computer-oriented
mathematics, i.e., discrete mathematics, algorithms, etc., in
mathematical education.

Furthermore, as the capabilities of computers increase, a question
arises: 2) is it possible to reduce or omit the part dealing with
topics that computers can do?

Let us consider the first question: Should the teaching of computer-
oriented mathematics be increased?

The mathematics used in computers is based on discrete and finite
numbers. In order for computers to advance further, a larger number of
scientists and engineers who have mastered such computer-oriented
mathematics will be needed. Therefore, it is obvious that the teaching
of computer-oriented mathematics will have to be increased in computer
related fields.

On the other hand, as computers are used more extensively, opportunities for non-computer-specialists to use computers will greatly
increase. Then, a question arises: should the teaching of computer-
oriented mathematics also be increased for those people? To answer

this question, it is necessary to consider how computers will develop in the future.

One of the most difficult problems in computer science is the low productivity of computer software. Thus, it is first necessary to increase the number of software engineers to increase that productivity. The immediate solution to this demand is to increase the teaching of computer-oriented mathematics to those in non-computer-majors, so as to make them computer experts. This is essential to cope with the shortage of software engineers in the short term. However, is it still essential in the long term?

One of the major reasons for the low productivity of computer software is believed to be the immaturity of computers. Therefore, a lot of research has now been directed toward the development of a new type of computer system which can be programmed much more easily.

In the long term, the successful results of such research will greatly improve software productivity. This implies that computers will be used or programmed by those who do not know computers very well.

Therefore, it is concluded that, in the long term, a great part of computer oriented mathematics will not have to be included in the general mathematics curriculum as long as the above research is successful.

We shall now consider the second question: Is it possible to reduce or omit the part devoted to topics that computers can do?

The progress of computers and their capacity for computer algebra in particular, is astonishing. Therefore, there arises the question: How can computer algebra systems be introduced into mathematical education?

To answer this question, it is necessary to consider the basic objectives of mathematical education. At first glance, it seems that computers can replace computational skills. For instance, why not let the computer algebra systems calculate derivatives all the time? Why should students spend so much time practising tedious calculations?

However, the above idea contains a crucial problem. For instance, can a student really understand a topic without first making efforts to solve many exercises by hand? Obviously, the answer is 'No'. The reason is the same when students are not allowed to use electronic calculators when learning addition, subtraction, etc.

Then another question arises: Isn't there any way at all in which computer algebra systems can be effective in mathematical education? The answer is affirmative. Namely, when a student is learning a topic, there is the essential material which cannot be replaced by computers, and the inessential material which is necessary as a tool to understand the topic but may be replaced by computers. And it is to the latter that computer algebra systems can be effectively applied.

For example, when learning indefinite integrals, complicated expansion into partial fractions or tricky transformation of variables may be classified as inessential material. When a student is learning indefinite integrals, such material should have already been learnt. Thus, computers may be used for the inessential material.

Therefore, when considering the introduction of computer algebra systems to mathematical education, it is primarily necessary to make the boundary between the above two parts clear. Moreover it is necessary to examine the curriculum for possible changes, taking account of computer algebra systems. Finally, the development of a method- ology for the effective use of computer algebra in mathematical education and an understanding of the effects of this methodology are necessary. These points will become increasingly important as computers progress.

3 A NEW WAY OF TEACHING MATHEMATICS
We shall now propose a new type of mathematical education model incorporating computer algebra systems and discuss its advantages. The discussion assumes a situation in the near future in which every student has a small computer algebra system on his/her desk and can use it with the ease of present electronic calculators.

3.1 Method of teaching
In the new teaching method, the course for teaching a topic A is divided into 1) 'the basic course', and 2) 'the application course'. The two courses are taught in that order.

In the basic course, the fundamental concepts of A are taught in almost the same way as the conventional method. However, the method is different in the following respects:

i)   Longer time must be spent on teaching the more fundamental concepts or principles. Complicated problems in which the complexity is not so essential to A should be reduced.

ii) More CAI systems, e.g., those with graphic images, etc., which help students understand A better may be used under the supervision of the teacher.

iii) Computer algebra systems are not allowed for the essential part of A, which is currently taught, because such use will hinder the student from a deeper understanding of A.

iv) However, the systems may be used for the knowledge B, which has already been acquired, in order to solve a question in A. In this case, it must be ensured that B is inessential to A. This draws students' attention to the more essential parts of the new topic A.

After the basic course, the purpose of the application course is to
clarify the position of A by applying it to more complicated problems
and learning the relation of A to other topics, thereby reaching a
deeper understanding of A. During this course, students are allowed
to use computer systems not only for inessential parts such as B but
for the essential part of A. Before using the computer system, of
course, instructions must be given on how to use the system, and the
limits, advantages and disadvantages of the system must be made clear.

A model of problem solving by the new teaching method incorporating
computer algebra systems is shown in Fig. 3.1. The model is different
from the conventional ones which do not use computers in that computers
are used for numerical calculations and algebraic operations (4), and
graphic images and simulation of the obtained results (5).

This model features the following advantages.

(1) Students can solve problems more quickly and with less effort by
allowing computers to do the work for (4) and (5) in Fig. 3.1. If the
input to the system is correct, the results from the system are free
from the mistakes which students might make during tedious calculations
by hand.

(2) A student can more quickly verify his understanding of the problem,
his basic strategy, and his mathematical formulations. Thus, he can
attempt the problem again more easily, if there is any mistake. Thus,
students can focus their attention on more intellectual work, i.e.,
problem understanding, planning basic strategy, verifying the obtained
results, etc. In addition, since students can verify their ideas more
quickly, they are encouraged to study further. This increases
students' incentive to study.

(3) Students can easily try several strategies for comparison. This
sort of learning leads to a development in students' proficiency in
obtaining an optimum strategy.

(4) Students can solve more problems in a limited time. Thus, they
can see the topic from a wider viewpoint, leading to a deeper understanding of the topic.

(5) This method does not hinder students from developing mathematical
thought. What is replaced by the computer has, in essence, little
relation to the development of it.

There is a point which must be noted when applying this new teaching
method. Since the computer systems return wrong answers to incorrect
inputs, it is extremely important to instruct students not to believe
the answers from computers absolutely. Therefore, greater efforts
must be made to develop proficiency, in order to be able to verify the
validity of the obtained results and select the right answer from a
number of outputs from the computer.

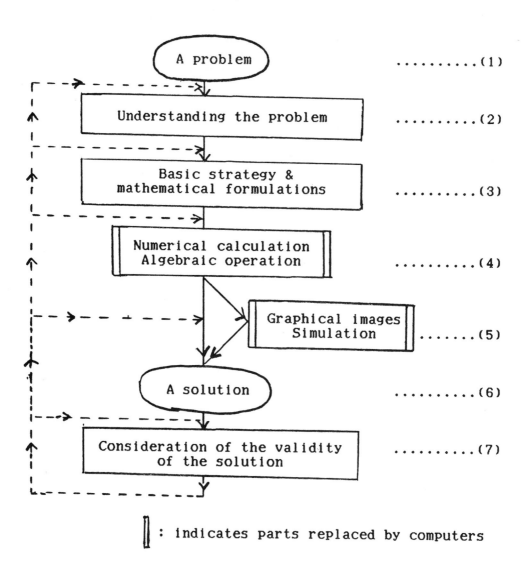

A problem .... (1)

Understanding the problem .... (2)

Basic strategy & mathematical formulations .... (3)

Numerical calculation Algebraic operation .... (4)

Graphical images Simulation .... (5)

A solution .... (6)

Consideration of the validity of the solution .... (7)

▯ : indicates parts replaced by computers

Fig. 3.1   A model of problem solving by the new teaching
            method incorporating computer algebra systems.
            (As indicated by dashed lines, it is possible
            to return from one point to any other.)

In the summary, the advantage of the new method is that the method can shift the focal point of mathematical education to more essential points, such as more emphasis on problem understanding, elaborating basic strategies and mathematical formulations and verifications of obtained results. Accordingly, a greater amount of more essential materials will be included in the mathematical curriculum.

In order to make full use of all the advantages, there are several points which computer algebra systems must feature.

(i)   The final output from the computer system is not always the most suitable answer for the students' use. Therefore, the systems must allow students to see the important intermediate results.

(ii)   The system should feature the following two operating modes.

> (1) Calculator mode: returns only final results. Students use the system just as a computational tool.

> (2) Trace mode: provides not only final results but major intermediate results, explanations of the rules used to reach the final results, etc. This mode allows students to under-stand the system which is usually treated as a black box, or to use the intermediate results.

(iii) A graphics system allowing results to be displayed from the algebra system or to be simulated must be effectively connected with a computer algebra system, allowing students to tackle problems more easily.

(iv)   The system must allow students numerical calculation as well as algebraic operations. Then, students can perform a numerical analysis of an expression obtained by the algebra system.

3.2 Feasibility of exploratory mathematics
Exploratory mathematics is a heuristic educational method which allows students experimentally or inductively to discover rules or theorems by themselves using computer mathematics systems. In this way, rules or formulas are not taught top-down, but bottom-up. Students use the computer to find rules and make hypotheses. Then, the students attempt to prove their hypotheses. A model for exploratory mathematics is shown in Fig. 3.2.

As an example, let us consider a case where the binomial theorem is to be taught. Before showing the expansion formula of $(1+x)^n$, students use the computer algebra system and expand the expression for $n = 0,1,2,3,....$.   Observing the obtained Pascal's triangle, the students can find the rule.

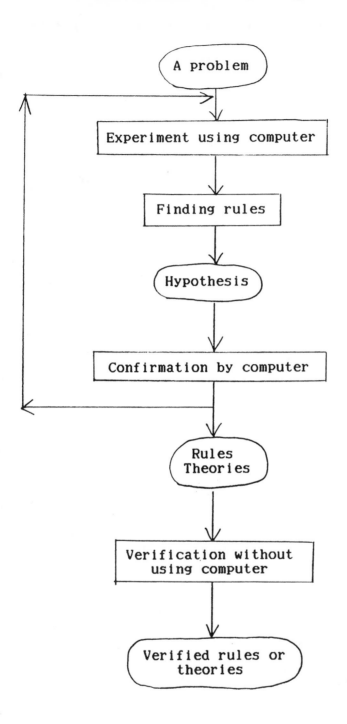

Fig. 3.2   A model of exploratory mathematics.

In addition, it may be expected that they find associated formulas at the same time, e.g.,

$$\binom{m}{n} = \binom{m-1}{n} + \binom{m-1}{n-1} .$$

This method would be impractical without computer algebra systems, because it would take too much time and effort. But, with computer algebra systems, this method of learning will give students pleasure when they discover something, promoting their incentive to study. Furthermore, the heuristic ability is extremely important not only for mathematics but for any scientific research. It should be stressed that such an ability can be developed by exploratory mathematics. In addition, this method can be used in the educational method stated in section 3.1. That is, during the basic course, this heuristic method can be included to teach specific topics.

Finally, there are topics in which exploratory mathematics is especially effective. Therefore, it is necessary to select those topics which are most suitable for this method.

4 CONCLUSIONS

We considered the influence of the progress of computers on mathematical education and proposed a new teaching method using computer algebra systems. What was shown is as follows:

1) There will be two types of changes brought about by computers; changes in the methodology and changes in the topics taught.

In the case of the former, the extensive use of CAI systems will inevitably be influential.

As for the latter, we considered the need for computer-oriented mathematics as well as the educational policy for the material which can be handled by computers. We showed:

i) In the short term, computer-oriented mathematics must be increased.

ii) In the long term, an increase in computer-oriented mathematics in the general mathematics curriculum will not be necessary due to the advance of computer technology.

With regard to educational policy, a new educational model which extensively uses computer algebra systems was proposed, and the advantages of it were considered.

2) The advantages of the new educational model are:

i) Students do not have to spend as much time as before on tedious calculations.

ii) Instead, they can concentrate on more essential and intellectual matters, i.e., problem understanding, elaborating basic strategies and mathematical formulations, and verifying the obtained results.

3) Exploratory mathematics can be exploited in the actual educational environment by the use of computer algebra systems. This method is especially effective in developing an ability for heuristics which is very important for all scientific work.

4) A revision of the curriculum will be necessary to incorporate computers into mathematical education.

Like it or not, the extensive use of computers is bringing about a variety of changes in our society and our daily lives. It is remarkable to note that computers can be an effective tool in attaining the ultimate goal of education, if the goal can be stated as to make a man who thinks by himself, who studies by himself, and who, having set himself questions, can think of the way to solve them. In this sense, the establishment of a most effective way of using computers in mathematical education is urgently needed.

REFERENCES

1   Loos, R., (1982). "Introduction", in Ed. B. Buchberger et al., Computer Algebra: Symbolic and Algebraic Computation, Springer Verlag, pp.1-10.
2   Hulzen, J. and Calmet, J., (1982). "Computer Algebra Systems", in Ed. B. Buchberger et al., Computer Algebra: Symbolic and Algebraic Computation, Springer Verlag, pp.221-243.

A FUNDAMENTAL COURSE IN HIGHER MATHEMATICS INCORPORATING
DISCRETE AND CONTINUOUS THEMES

S. B. Seidman
George Mason University, Fairfax, Va. 22030, USA

M. D. Rice
George Mason University, Fairfax, Va. 22030, USA

THE CURRENT MATHEMATICS CURRICULUM: TRADITIONS AND CONCERNS

For many years, a crucial place in the mathematics curriculum of the last year of secondary school or the first year of
university studies has been occupied by the differential and integral
calculus. The calculus can be seen both as the culmination of the
secondary school mathematics curriculum and as the beginning of the
serious study of mathematics in the university. In some sense, the
study of calculus has become synonymous with the serious study of mathematics. The central and essential position occupied by calculus can
be traced to at least two interrelated causes.

For mathematicians, calculus represents the methodology and techniques
needed for the study of functions, first defined on the real line, then
on higher-dimensional Euclidean spaces, and finally on the complex
plane. Thus, the study of the calculus allows students for the first
time to acquire the formal and abstract tools that are essential for
the further study of higher mathematics.

On the other hand, calculus provides the foundation for the application
of mathematics to the physical sciences and engineering. These applications date back to Newton's original development of the calculus in the
seventeenth century, and since that time they have been wildly successful across a vast collection of disciplines, even including (in recent
years), the biological sciences and economics. All of the calculus-
based applications are based on mathematical models that can be regarded as being continuous; that is, the quantities being modeled are
real numbers (or elements of some Euclidean space R^n).

Given both the central mathematical position of the calculus and its
vital role in applications (not to speak of the interaction between
these two features), it is easy to see why the calculus has occupied
such a fundamental and unassailable position in mathematics curricula.
During the past several decades, however, the central role of calculus
has been seriously questioned, and the questions have been repeated
with particular emphasis during the last few years (Ralston 1981).
Just as a major motivation for the predominance of calculus in the
curriculum has been the wide range of the applications of continuous
mathematics, the challenge to that predominance has arisen from the
steadily increasing interest in the applications of discrete mathematics
in many disciplines.

This increasing interest in discrete mathematical applications can be primarily attributed to the widespread use of computers. Computers are essentially discrete machines, and the mathematics that is needed to describe their functions and develop the algorithms and software needed to use them is also discrete. As a consequence, the discipline of computer science is heavily dependent on a wide variety of discrete mathematical ideas and techniques. Furthermore, the easy availability of computers has encouraged the use and development of discrete mathematical models in many disciplines. For one example, operations research models (linear programming, integer programming, etc.) are widely used and are based on a discrete mathematical perspective.

It is natural to expect that the rapid growth of interest in discrete mathematics and its applications, fueled by the explosive developments associated with computers, should have an impact on the mathematics curriculum. Although this impact would have been significant under any circumstances, its effect has been magnified by other questions that have been raised in the United States in recent years about the teaching of calculus. Widespread dissatisfaction has been reported with the nature of the calculus courses and the knowledge of the students that have completed them (Lochhead 1983, Steen 1983). The computer is also directly influencing the content of the calculus course itself, both by encouraging the inclusion of numerical methods and by suggesting that symbolic manipulation software may make emphasis on techniques of differentiation and integration obsolete (Bushaw 1983, Wilf 1983).

In summary, both the nature of the calculus course and the fundamental position that calculus has occupied in the mathematics curriculum for more than a century have come under serious challenge. These challenges have come both from within and outside the community of mathematicians, and they can primarily be attributed to the increasingly broad role that computers are playing in the various scholarly disciplines represented within the university and in the wider world. In the next section of this paper, we will look at the responses that have been proposed to these challenges.

RESPONSES TO THE CHALLENGE OF DISCRETE MATHEMATICS
When any curriculum is confronted by a new topic that should be included, there are essentially two potential responses. The new topic can either be encapsulated in a course that is added to the curriculum, or it can be incorporated as a fundamental constituent of a revised curriculum. Most topics that have been added to the mathematics curriculum in recent decades have been added as new courses (e.g. abstract algebra and topology).

It was therefore natural that when mathematics faculties were asked to include discrete mathematics in the curriculum, this was most commonly done by developing new courses in discrete mathematics. Such courses were designed primarily for students of computer science. There were

two fundamental problems with this approach. First, the discrete mathematics courses were usually taken by third-year students, so that the
material was learned too late to be of use in the data structures
courses taken by first and second year students of computer science.
Second, when students were expected to use their discrete and continuous mathematical skills in fourth-year computer science courses (for
example, in the analysis of algorithms), most have found it very difficult to combine these skills effectively. Many students do not see any
connections between discrete and continuous mathematics, and are unable,
for example, to apply calculus techniques to estimate growth rates of
discrete functions or to estimate the size of discrete sums. This in-
ability to combine discrete and continuous skills is also found in students of probability, operations research and signal processing.

Both of the above reasons suggest that discrete mathematics should be
incorporated as a component of the fundamental mathematics course that
is offered to all students in their first two years of university
study. This suggestion was first made by Ralston (1981), who proposed
that the study of discrete mathematics precede the study of calculus.
He argues that such an organization would benefit virtually all students of mathematics, and not just those students concentrating in computer science. Ralston's proposal has led to substantial discussion in
the United States on the proper place of discrete mathematics in the
curriculum (Ralston & Young 1983). The debate has focused on whether
discrete mathematics should precede or follow the calculus in the curriculum of the first two years. Many of the arguments advanced on
either side are administrative in nature, dealing either with the demands of other curricula (such as physics or engineering) or with
articulation with other institutions (such as high schools, junior
colleges or universities that have retained the standard curriculum).

Whether calculus is placed before or after discrete mathematics, it is
by no means clear that students who have completed both courses will be
able to combine their discrete and continuous mathematical skills in an
effective manner. This problem has been recognized by some designers
of proposed curricula, and consequently their calculus proposals generally include some discrete aspects, such as extended discussion of
numerical methods and substantial use of sequences (see for example
Bushaw 1983).

Another possibility, which is rarely given serious attention, would be
to develop a new, unified curriculum that would interweave discrete and
continuous themes throughout its courses. While the first year of the
curriculum would correspond to the calculus course, its real thrust
would be the study of functional behavior and functional representation. The course would consider discrete functions (sequences) along
with continuous functions, and would constantly emphasize analogies and
parallels between discrete and continuous situations. Thus the first
year of the curriculum would be primarily continuous, but with a strong
discrete flavor. The second year of the curriculum would focus on
structure, and would be primarily discrete, but with a strong continuous flavor.

This paper will argue that a curriculum unifying discrete and continuous themes is not only feasible, but has the potential of providing
students with a broad, powerful perspective embracing the mathematical
ideas and techniques that are needed for the study of computer science.
This perspective would also yield a strong mathematical foundation for
the study of engineering, the physical sciences, and indeed for the
study of higher mathematics itself.

Furthermore, the development of such a curriculum will force a reexami-
nation of the topics taught in the conventional calculus course. As
mentioned above, various recommendations have been made to remove or
include particular topics. Although each such recommendation has been
solidly grounded, no consistent rationale has been given for the collection of topics that together make up the proposed calculus course. The
first-year course outlined below has a consistent theme - functional
behavior and representation - and each topic to be included in (or ex-
cluded from) the course should be judged on the degree that it matches
the course's perspective.

In the following section, a detailed outline and discussion will be
given only for the first year of the proposed two-year curriculum. At
the conclusion of the paper, we will return to the second year of the
curriculum, as well as to the larger issues raised by the question of
articulation with other curricula.

A FIRST-YEAR CURRICULUM INCORPORATING DISCRETE AND CONTINUOUS THEMES

The fundamental thrust of the proposed first-year curriculum is the behavior and representation of functions. Roughly, the
first semester is devoted to tools for the description and analysis of
functional behavior, with the focus shifting to representation of functions in the second semester. Before presenting a more extended discussion of the benefits to be achieved by including both discrete and
continuous topics, it will be useful to give an annotated outline of
the first semester curriculum.

A. Functions

1. Numbers and Relations

A knowledge of set concepts and notation is assumed. Inequalities will be emphasized.

2. Functions and Operations

Function concept and functional notation will be
introduced, stressing the algorithmic interpretation of the function symbol f. Discussion will
include domain and range, operations on functions

(arithmetic operations, composition, translation),
and graphs of functions. Useful functions will be
introduced (polynomials, rational functions, exponential functions (defined on the integers), abso-
lute value, floor, ceiling).

3. Models

Algorithms and elementary complexity analysis will
be introduced (including binary search). This
will allow discussion of the function[lg(n)].
Models demonstrating the need to construct functions and to perform curve fitting will be included.

B. Behavior of discrete functions

1. Sequences: Iteration and Recursion

This section will include a discussion of geometric series. Examples will include the
Fibonacci numbers and the greatest common divisor
function.

2. Difference Operators

The difference operator Δ will be introduced as a
function on sequences. The recursion scheme

$$u_{k+1} - u_k = \Delta u_k$$

will be treated in order to emphasize the special
functions (defined on the integers). Formulas for
higher differences will be discussed.

3. Summation

The primary topic here will be the binomial theorem,
both in its standard form and in the expression for

$$(1 + \Delta)^n.$$

The second form will allow various formulas for
finite sums to be presented.

4. Landau Notation (O,o) and Limits of Sequences

C. Behavior of continuous functions

1. Limit Heuristics

Limits of functions will be discussed only in

terms of limits of sequences. The continuity concept will be introduced. The operator

$$\Delta_h f = (f(x+h)-f(x))/h$$

will be introduced. Analogies to the discrete
difference operator discussed above will be pursued.

2. First Derivative

The derivative will be defined, and interpreted
using tangent lines. It will be shown that differentiable functions are continuous.

3. Differentiation Rules

Powers and roots; product and quotient rules.

4. Monotone Functions and Local Extrema

A rigorous treatment will be postponed. Curve
sketching will be introduced here.

5. Second Derivative

Concavity will be discussed and applied to curve
sketching.

6. Extreme Values

Maximum-minimum problems will be solved. Examples
will also demonstrate the use of piecewise linear
functions.

7. Related Rates

The chain rule will be presented, and related rate
problems will be solved.

D. Estimation and error

1. Mean Value Theorem

Monotone functions will be discussed more rigorous-
ly, and the MVT will be applied to global estimation of functions.

2. Solution of Equations

Newton's method will be discussed from both geometric and iterative perspectives. An elementary

treatment of error estimation will be given, and critical values will also be estimated.

3. Interpolation

Interpolation of functions by lines and parabolas will be discussed, using the difference operators developed above.

4. Approximation

Second-order Taylor polynomials will be used to approximate functions, and the estimated error will be computed. Analogies will be drawn between interpolation and approximation and between differences and derivatives.

E. Integration

1. Introduction

The summation operator for sequences will be introduced. Its relation to the difference operator will be discussed. It will be treated as an aggregation operator, and used to motivate the discussion of area.

2. The Definite Integral

This will first be introduced using a piecewise linear definition. This definition will then be applied to step functions. The area definition will then be presented, and applied to parabolas using the results on finite sums obtained above. Some elementary properties of the definite integral will be presented, including the mean value theorem for definite integrals.

3. The Indefinite Integral

This will be explicitly computed for step functions, piecewise linear functions and parabolas.

4. The Fundamental Theorem of Calculus

This will be derived from the mean value theorem for definite integrals. The chain rule will be applied to investigate some properties of the integral of $1/x$.

5. Evaluation of Integrals: Analytic Techniques

Substitution techniques will be discussed, as well
as the use of integral tables.

6. Evaluation of Integrals: Numerical Techniques

The trapezoidal rule and Simpson's rule will be
discussed. It will also be shown how integrals
can be estimated using inequalities, and how sums
can be estimated using integrals.

7. Applications of Integration: Aggregation

The applications to be treated include work and
volume.

8. Applications of Integration: Modeling

The primary theme here will be the recognition of
Riemann sums in differing situations. Examples
will be taken from arclength and fluid flow. The
basic point will be that a model generates a discrete (Riemann) sum, which can then be approximated
by a definite integral.

Although this annotated outline gives a good overview of the first semester of the proposed course, it is too brief to show how the inter-
weaving of discrete and continuous themes can lead to major benefits.
The following examples are meant to be typical of the perspective that
will be possible within this course structure.

Example 1: At the beginning of the course, the discrete exponential
function, $f(n) = 2^n$, will be introduced, along with its one-sided inverse, $g(n) = \max\{k \mid 2^k \leq n\}$. The function $g(n)$ is vitally important
in computer science; for example, $g(n)+1$ is the worst-case number of
comparisons in a binary search of a list of length n. The growth rate
of $g(n)$ is important, and is usually treated (via calculus) using
L'Hospital's rule. We suggest a discrete approach, based on the
binomial theorem. Clearly $2^{g(n)} \leq n$, so that $g(n)/n \leq g(n)/2^{g(n)}$, and
$g(n)$ approaches ∞ with n since $g(2^L) = L$. Thus it is only necessary
to look at the behavior of $k/2^k$ as $k \to \infty$. By the binomial theorem
$2^k = (1+1)^k \geq k(k-1)/2$, and hence $k/2^k \leq 2k/k(k-1) = 2/(k-1)$, which
gives the desired result. The simplicity of the discrete argument
should aid the student in learning, understanding and assimilating the
growth rate of the <u>continuous</u> logarithm.

Example 2: The syllabus outline has referred to analogies between the
discrete difference and summation operators on the one hand, and differentiation and integration on the other. For example, the difference
operator is defined on the sequence $\{u_n\}$ by $\Delta u_n = u_{n+1} - u_n$. If we
define a function on the integers by $x^{(m)} = x(x-1)\ldots(x-m+1)$, then it
is easy to see that $\Delta x^{(m)} = mx^{(m-1)}$, $\Delta^2 x^{(m)} = m(m-1)x^{(m-2)}$, and
finally that $\Delta^m x^{(m)} = m!$ and $\Delta^{m+1} x^{(m)} = 0$. Thus the behavior of the

difference operator (and its iterates) on the polynomials $\{x^{(m)}\}$ is strongly analogous to the behavior of the differentiation operator (and its iterates) on the polynomials $\{x^n\}$. Furthermore, since each collection of polynomials provides a basis for the vector space of polynomials of degree at most n, an example has been introduced which will be useful in a later course in linear algebra.

One further benefit of the use of difference operators is the natural observation that $\Delta 2^n = 2^n$, or more generally that $\Delta k^n = (k-1)k^n$. This suggests that exponential functions, whether discrete or continuous, may have a special role to play with respect to difference or derivative operators, and serves to motivate the later observation that $d/dx(e^x) = e^x$.

Example 3: The first two examples used discrete ideas to motivate continuous concepts that are to be introduced later. In this example, continuous techniques are used to obtain a discrete result. The identity giving the sum of a geometric progression,

$$\sum_{k=0}^{n-1} x^k = (x^n-1)/(x-1)$$

can be differentiated using the quotient rule to obtain the identity.

$$\sum_{k=1}^{n-1} kx^k = ((n-1)x^{n+1} - nx^n + x)/(x-1)^2.$$ Using this identity, it is

immediate that $\sum_{k=1}^{n-1} k2^k = (n-2)2^n+2$ and that $\sum_{k=1}^{n-1} k2^{-k}=2 - (n+1)/2^{n-1}$.

The last result yields $\sum_{k=1}^{\infty} k2^{-k} = 2$, since it has already been shown

that $k/2^k \rightarrow 0$ as $k \rightarrow \infty$. This example serves to remind students that continuous techniques can be important in discrete situations.

These examples demonstrate that the proposed course does not merely insert a collection of important discrete topics into the calculus course, but rather expresses a consistent approach to all of the subject matter. The fundamental perspective is the study of functional behavior, and both discrete and continuous functions are treated throughout. Each class of functions is used to develop tools and suggest analogies that will be useful for the study of functions of the other class.

The second semester of the course further elaborates its functional perspective. Rather than give a detailed, annotated outline, we will discuss the topics to be covered and describe how they relate to the themes developed during the first semester. The second semester is primarily devoted to material taken from two broad categories: special functions and representation of functions.

Exponential and logarithmic functions will be treated in depth. The natural logarithm will be introduced using the definite integral, and

its properties will be investigated. The inverse of the logarithm will
be motivated using growth models and the differential equation
dy/dx = ky, and the relationship of this inverse to the exponential
function will be motivated using difference equations and the discrete
logarithm. Finally, the properties of the function e^x will be developed. Numerical estimates for exponential and logarithmic functions
will be used throughout the discussion.

The next major topic will be trigonometric functions. Here the primary
motivation will come from the geometry of the circle and from models of
circular and harmonic motion, although discrete periodic functions, such
as mod n, will also be used. The properties of the trigonometric
functions will be developed. Integration by parts will be introduced
and applied to the special functions. The special integrals leading to
the inverse trigonometric functions will be introduced here. Mathematical models suggesting the use of trigonometric polynomials will
also be used.

Once the special functions have been treated, it will be natural to
discuss various forms of infinitary behavior. The discussion will
begin with a reconsideration of infinite sequences, including a presentation of indeterminate forms and their applications to Landau
notation. The remainder of this section will be devoted to improper
integrals and infinite series, emphasizing the analogies between these
two forms of infinite summation.

At this point, the focus will shift somewhat from functional behavior
to functional approximation and representation. Thus the next major
topic will be power series, with particular emphasis on the use of
Taylor series to represent functions. Generating functions for simple
recursions will be discussed, and a certain amount of attention will be
devoted to computational issues and the estimation of error terms. The
constant theme will be the use of Taylor series as function approximations to obtain information about functional behavior that would
otherwise be difficult to obtain.

The final topic will be trigonometric series, with particular emphasis
on the representation of functions using Fourier series. The treatment
of Fourier series at this early point will require the introduction of
complex numbers, which will reinforce the students' geometric understanding of trigonometric functions. Furthermore, the availability of
Taylor series will permit an analytic as well as a geometric discussion
of the identity e^{ix} = cos x + i sin x. Finally, the early introduction
of Fourier series will make it possible to discuss discrete Fourier
series and their applications at a far earlier point in the curriculum
than is presently possible.

Clearly, the focus on functional behavior and representation has produced a first-year course that is rather different from what is currently taught. The essential core of the current calculus course has
been retained, but it is always made clear that is is there because it
throws a powerful spotlight on functional behavior and representation.

Conversely, many traditionally taught topics have been removed. This pruning was only possible because the developers approached each topic with the same question: how does this topic impact on the main theme of the course?

Now that the course has been outlined, it remains to be seen how it will fit into the curriculum. We will also have to pay some attention to the second-year course that will follow this course, and also to the political and institutional problems that its adoption would pose.

IMPLICATIONS FOR THE CURRICULUM

The first question to be addressed is the audience to be served by the proposed course. It is clearly ideally suited for students of computer science, since it merges themes from continuous and discrete mathematics in a synergistic manner. Students who have successfully completed the course can be expected to handle the mathematics arising (for example) in the analysis of algorithms. It can also be argued that this course would be well suited as a first course for students of mathematics, the physical sciences and engineering. For these disciplines, the major omission has been vector geometry and multivariate calculus. In many universities, a large proportion of this material is treated in the second year, and it is not unreasonable to suppose that even more could be shifted to a third-semester course designed for those students.

Although much vitally important mathematics can be subsumed under the general heading of "functions", an equally important heading is that of "structure". While the proposed course is intended to give students the most important tools that come under the former heading, it does not address the latter. For students of computer science, both headings are equally important, and thus an important place in their education must be found for "structure". Much of the debate summarized above on the place of discrete mathematics in the curriculum can be seen as a debate on the place of "structure" in the curriculum. Following on the first-year course that has been outlined above, it is reasonable to believe that a second-year course focusing on "structure" can be developed.

Such a course will not be described here, but it is possible to discuss briefly what general topics might be included. The primary strands might be discrete mathematics, linear algebra and probability theory. Discrete mathematical topics could include relations, graphs, Boolean algebras and formal languages. The discussion of linear algebra could include some multivariate calculus, which could then be applied in the probability portion of the course. Just as with the first-year course, the topics included in the second-year course should be chosen because they illustrate vital structural themes or because they are motivated by or permit the development of important applications.

The introduction of courses designed along these lines will not be a simple matter. The obstacles that will be found will range from the need for new textual materials to the difficulty of articulating the new courses with other institutions on all levels. It would be an unfortunate mistake, however, to conclude that because of the certainty of encountering what seem to be insuperable obstacles to the introduction of a truly new curriculum, the only possible strategy is one of incremental change. The development and introduction of a curriculum integrating discrete and continuous ideas is an exciting challenge, and one that is sure to be taken up in several places. What is really needed is a collection of design and development experiments, performed in out-of-the-way "protected" environments. Once a new curriculum has proven its viability and worth in one or more of these experimental environments, it will be time to address the structural and institutional issues involved in transplanting the successful curriculum to less protected situations.

REFERENCES

Bushaw, D. (1983). 'A two-year lower-division mathematics sequence'. In The Future of College Mathematics, eds. A. Ralston & G.S. Young, pp. 111-118. New York: Springer-Verlag.

Lochhead, J. (1983). 'The mathematical needs of students in the physical sciences'. In The Future of College Mathematics, eds. A. Ralston & G.S. Young, pp. 55-69. New York: Springer-Verlag.

Ralston, A. (1981). 'Computer sciences, mathematics and the undergraduate curriculum in both'. Amer. Math. Monthly 88, 472-485.

Ralston, A. & G.S. Young, eds. (1983). The Future of College Mathematics. New York: Springer-Verlag.

Steen, L.A. (1983). 'Developing mathematical maturity'. In The Future of College Mathematics, eds. A. Ralston & G.S. Young, pp. 99-107. New York: Springer-Verlag.

Wilf, H.S. (1983). 'Symbolic manipulation and algorithms in the curriculum of the first two years'. In The Future of College Mathematics, eds. A. Ralston & G.S. Young, pp. 27-40. New York: Springer-Verlag.

GRAPHIC INSIGHT INTO CALCULUS AND DIFFERENTIAL EQUATIONS

David Tall
Mathematics Education Research Centre
Warwick University, COVENTRY CV4 7AL, England

Beverly West
Mathematics Department, Cornell University,
Ithaca, NY 14583, U.S.A.

The human brain is powerfully equipped to process visual information. By using computer graphics it is possible to tap this power to help students gain a greater understanding of many mathematical concepts. Furthermore, *dynamic* representations of mathematical processes furnish a degree of psychological reality that enables the mind to manipulate them in a far more fruitful way than could ever be achieved starting from static text and pictures in a book. Add to this the possibility of student exploration using prepared software and the sum total is a potent new force in the mathematics curriculum.

In this paper we report on the development of interactive high resolution graphics approaches at different levels of teaching calculus and differential equations. The first author has been concentrating on the calculus in the U.K. [Tall 1985] and the second is working with John H. Hubbard in the U.S.A. on differential equations [Hubbard & West 1985], (later referred to as [T] and [H&W] respectively). We are particularly grateful to Professor Hubbard for his assistance in the preparation of this article.

Others have pioneered a computer approach to these topics, particularly [Artigue & Gautheron 1983] who used computer graphics to build up pictures of solutions of autonomous systems of differential equations and [Sanchez et al. 1983] who emphasized a qualitative approach to the theory. A suitable qualitative approach can lean to an insightful understanding of the formal quantitative theory. The major advance in our work is the interactive nature of the prepared software, enabling students to explore the ideas and develop their own conceptualizations.

DIFFERENTIATION
Traditionally the notion of differentiation is founded on the idea of a limit, either geometrically as a chord approaches a tangent, or algebraically as a ratio $(f(x+h)-f(x))/h$ as h tends to zero.

The computer brings new possibilities to the fore; we may begin by considering the gradient not of the tangent, but of the graph itself. Although a graph may be curved, under high magnification a small part of it may well look almost straight. In such a case we may speak of the gradient of the graph as being the gradient of this magnified

(approximately straight) portion. A tiny segment of the graph $y=x^2$
near x=1 magnifies to a line segment of gradient 2 (figure 1).

To represent the changing gradient of a graph, it is a simple matter to
calculate the expression $(f(x+c)-f(x))/c$ for a small fixed value of c as
x varies. As the chord clicks along the graph for increasing values of
x, the numerical value of the gradient of each successive chord can be
plotted as a point and the points outline the graph of the gradient
function (figure 2). In this case the chord gradient function of sin x
for small c approximates to cos x , which may be checked by superimposing
the graph of the latter for comparison. Thus the gradient of the graph
may be investigated experimentally before any of the traditional
formalities of limiting processes are introduced.

Figure 1

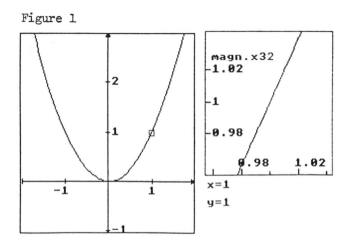

Figure 2

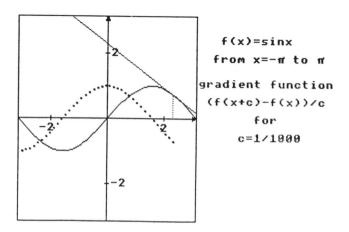

The moving graphics also enable the student to get a dynamic idea of
the changing gradient. Students following this approach can see the
gradient as a global function, not simply something calculated at each
individual point.

When the standard formulae for differentiation are developed, the
symbols dx, dy can be given a meaning as the increments in x,y to the
tangent. Better still, (dx,dy) may be viewed as the *tangent vector*, a
valuable idea when we come to the meaning of differential equations.

NON-DIFFERENTIABLE FUNCTIONS
If one views a differentiable function as a "locally straight" graph,
then it is easy to explain the notion of non-differentiability. The
graphs of $|x - 1|$ at x=1 or $|\sin x|$ at multiples of π clearly do not
magnify to look straight. At these points they have different left and
right gradients which magnify to give half-lines meeting at an angle.

The function of [Takagi 1903] may be drawn by a computer program. It is
built up in stages starting from the saw-tooth y=s(x) given by taking
the decimal part d=x-INTx of x and defining

$$s(x)=d \text{ if } d < \tfrac{1}{2}, \text{ otherwise } s(x)=1-d.$$

The sequence of functions

$$b_1(x) = s(x)$$
$$b_2(x) = s(x) + s(2x)/2$$
$$\ldots$$
$$b_n(x) = s(x) + \ldots + s(2^{n-1}x)/2^{n-1}$$

tends to the Takagi function (figure 3).

Figure 3

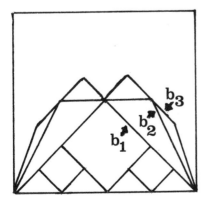 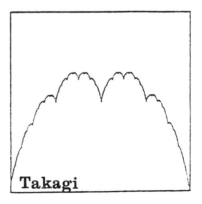

The process may be drawn dynamically on a VDU; we regret it cannot be pictured satisfactorily in a book, not even this one. But higher magnification of the Takagi function using prepared software shows it nowhere magnifies to look straight, so it is nowhere differentiable. This intuitive approach can easily be transformed into a formal proof of disarming simplicity [Tall 1982].

AREA CALCULATIONS

Computer graphics can be used to draw the pictures of area approximations as they are being calculated. For example [T] computes the area under a graph by a variety of rules (first ordinate, last ordinate, midordinate, trapezium or Simpson) and displays the areas calculated in different colours according to the sign. If the calculation has a positive step, the picture draws successive areas from left to right with positive ordinate giving a positive area and negative ordinate giving a negative area (figure 4). A negative step builds up the picture from right to left and it is equally easy to see that this reverses the signs; a concept traditionally found difficult before the advent of moving graphics.

These graphic facilities are not just introductory material for beginners. They can be used for visualizations of subtle theorems. Take the case of a discontinuous (Riemann integrable) function whose area function is not differentiable where the original function is discontinuous. When the cumulative area function for f(x)=x-INTx is drawn as a sequence of dots, the area graph is visibly continuous, but it has "corners" where the original function has discontinuities (figure 5).

Figures 4 & 5

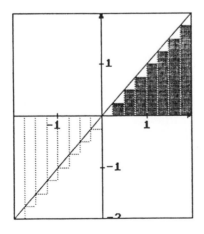
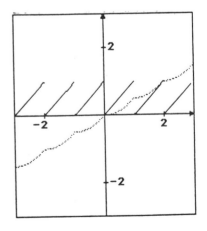

ANTIDIFFERENTIATION

This is usually regarded as reversing the formulae for integration, but
it may also be characterized graphically as knowing the gradient
dy/dx=f(x) and requiring a graph y=I(x) with this gradient. This
information may be represented graphically by drawing an array of short
line segments through points (x,y) with gradient f(x). A solution
y=I(x) is traced out by following the direction lines (figure 6).
It satisfies I'(x)=f(x).

As the gradient direction is a function of x alone, the solution curves
differ by a constant. If these are drawn numerically using a constant
step along the graph, rather than a fixed x-step, the solution in simple
cases will remain on a connected component of the graph. For instance,
a solution curve of dy/dx=1/x starting to the right of the origin always
remains on the right. Thus two different antiderivatives must differ by
a constant only over a *connected component* of the domain. The role of
the "arbitrary constant" is seen in its true light.

FIRST ORDER DIFFERENTIAL EQUATIONS

In graphical terms, the solution of a first order differential equation

$$dy/dx = f(x,y)$$

is simply an extension of the antidifferentiation idea: draw a
direction diagram and trace a solution by following the given directions.

Figure 6

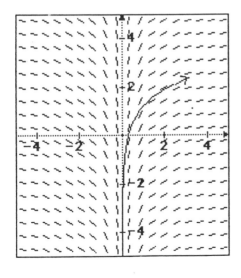

If one takes the equation

$$y\frac{dy}{dx} = -x$$

then there are no global solutions of the form y=f(x); a solution is an *implicit* function

$$x^2+y^2 = \text{constant}$$

which is a circle centre the origin. The "first order differential equation" program in [T] has a routine to follow round implicit flow lines (figure 7). At points where the flow-lines meet the x-axis the tangents are vertical and the interpretation of dy/dx as a real function fails, but the vector direction (dx,dy) is valid with dx=0 and dy non-zero. Thus a first order differential equation is sometimes better viewed as giving implicit information about the direction of the tangent (dx,dy) rather than explicit information about the derivative.

The Cornell program DIFFEQ 2 in [H&W] uses the same approach, allowing students to get qualitative feelings for all the solutions at once and relating them meaningfully to the theory (figure 8).

A combination of numerical methods and pictures can give enormous insight to standard formal approaches, often cruelly exposing the limitations and downright misrepresentations found in many elementary texts.

Figure 7

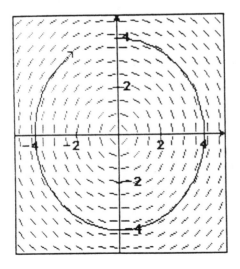

EXISTENCE OF SOLUTIONS TO DIFFERENTIAL EQUATIONS
There comes a time in a university course on differential equations when
honesty requires the teacher to admit that the (cookbook) methods for
solving differential equations usually fail. Such innocent looking
equations such as

$$dy/dx = y^2 - x, \quad dy/dx = \sin(xy), \quad dy/dx = e^{xy}$$

do not have solutions that can be written in elementary terms.
Students often mistakenly confuse this with the idea that the equations
have no solutions at all. However, if they are able to interact with a
computer program that plots a direction field and then draws solutions
numerically following the direction lines, the phenomenon takes on a
genuine meaning: "Of course the equations have solutions: we can see
them". They are immediately drawn to analysing the solutions using the
computer in a way that was previously impossible.

QUALITATIVE ANALYSIS OF DIFFERENTIAL EQUATIONS
New forms of analysis emerge now we can see as many solutions as we wish
all at the same time. In figure 8, notice how the solutions tend to
"funnel" together moving to the lower right-hand side; in the upper
right they spray apart (an "antifunnel"). Qualitatively descriptive
terms such as "funnel" and "antifunnel" can be defined precisely to give
powerful theorems with accurate quantitative results [H&W].

For example, the equation $dy = y^2 - x$ in figure 8 has two overall
behaviours: solutions either approach vertical asymptotes for finite x

Figure 8

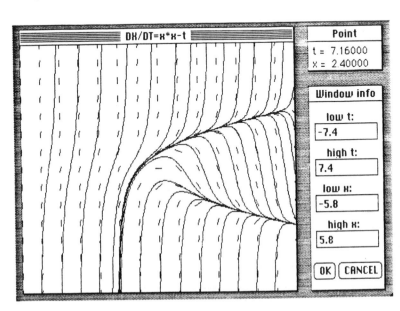

or fall into the funnel and approach $y = -\sqrt{x}$ as $x \to \infty$. In the antifunnel there is a unique solution approaching $y = +\sqrt{x}$ which separates the two usual behaviours. Furthermore, the qualitative techniques enable us to estimate the vertical asymptote for a solution through any given point with good precision.

NEWTON'S LAWS

The classical three-body problem defies elementary analysis, yet a computer program can cope with relative ease. The program PLANETS in [H&W] takes a configuration of up to ten bodies with specified mass, initial position and velocity and displays the movement under Newton's laws (figure 9). The data may be input either graphically with the cursor, or numerically in a table.

Figure 9

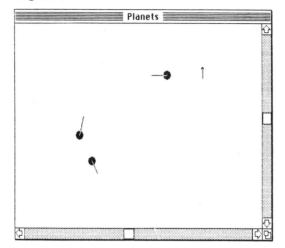

masses in position, with velocity vectors at time t_o

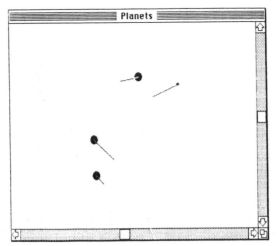

a little later under the action of Newton's Laws

The program allows exploration of possible planetary configurations and it soon becomes plain that stability is the exception rather than the rule. One may wonder under what circumstances stability occurs. Other questions arise, such as the braided rings of Saturn that were a great surprise when they were discovered by the Voyager space flight. Nobody imagined this kind of behaviour until it was observed, yet braided behaviour showed up in the very first experiments with the PLANETS program; we still don't know whether this is a common or an exceptional occurrence.

SYSTEMS OF DIFFERENTIAL EQUATIONS

For an autonomous system dx/dt=f(x,y), dy/dt=g(x,y), the computer can draw a direction field and trajectories in the x,y phase plane. The program SYSTEMS 2 in [H&W] will also locate singular points starting at any point in the field using Newton's method, also drawing separatrices for saddle points (figure 10).

The use of the computer permits study of individual differential equations and systems that are far too difficult to want to attack by hand. Furthermore, the dynamic interactive programs allow the user to see the direction and speed with which the solution moves, and to sense the stability of a limit cycle as the solution moves in.

Figure 10

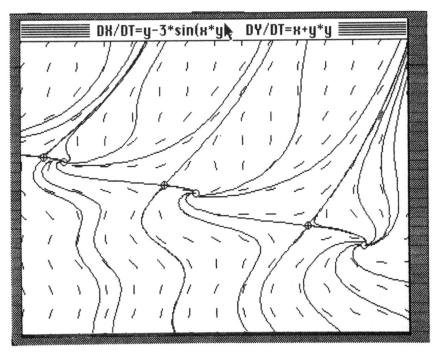

The use of the computer permits the study of individual differential equations and systems that are far too complicated to attack by hand. Figure 11 shows the solutions of the system

$$dx/dt = \cos(y), \quad dy/dt = \sin(xy)$$

and figure 12 shows the solution of the polar differential equations

$$dr/dt = \sin(r), \quad d\theta/dt = \cos(r)$$

which exhibits limit cycles for r=kπ. Both are drawn using the technique of [Artigue & Gautheron 1983].

Figure 11

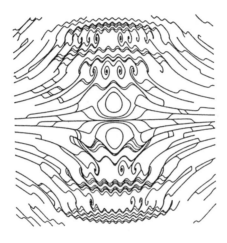

Figure 12

SECOND ORDER DIFFERENTIAL EQUATIONS
The simple direction field for first order equations appears to break
down in the second order case: a second order differential equation, say,

$$d^2y/dx^2 = -x$$

does not have a direction field in the x-y plane. There are an infinite
number of solutions through each point, one for each starting direction,
as may be investigated using the computer [T] (figure 13).

Solutions of such equations are often attacked by introducing a new
variable $v = dy/dx$, giving a (non-autonomous) system of two linear
equations:

$$dy/dx = v$$
$$dv/dx = -x.$$

In three-dimensional space, at every point (x,v,y) these equations give
the tangent vector (dx,dv,dy) in the direction $(1,v,-x)$. Thus there is
a direction field, but it is in three dimensional space, not two. [T]
draws a solution following this direction field, with the three -
dimensional solution simultaneously projected onto the two main
coordinate planes (x,y) and (x,y), representing the velocity v and the
distance y as x increases (figure 14). (The program also allows the
three-dimensional view to be replaced by the (y,v) phase-plane to give
all three coordinate planes simultaneously.) The understanding of the
nature of the solutions is greatly facilitated by seeing them evolve
dynamically in space, producing the set of solutions (figure 13) as the
projection of a three-dimensional picture onto the (x,y)-plane.

Figure 13

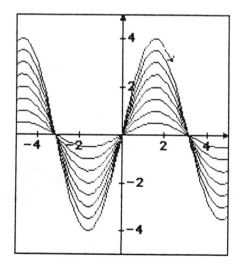

CHANGES IN LEARNING STYLE

The programs in [T] and [H&W] are powerful general purpose utilitities rather than self-contained programmed learning. Those in [T] were designed for teacher demonstration as well as student exploration, with facilities to slow down or stop the action to illustrate a point. After an introduction to the ideas, students may draw the gradient of $f(x)=x^n$ for $n = 1,2,3, \ldots$ to *investigate* the pattern and *conjecture* the formula for the derivative of x^n; they may *test* it for values such as $n = 4,5$ or $n = -1, -2, 1/2, \pi, \ldots$ before going on to *prove* the result for simple values of n. In drawing the graphs, students gain valuable appreciation of the range of values for which the formulae are valid, a factor often sadly lacking in blind algebraic manipulation.

In the calculus students may investigate the gradients of functions such as sine, cosine, tangent, exponential and logarithm, and conjecture the formulae before they are derived formally. In differential equations they may explore problems at the boundaries of research (such as the rings of Saturn) and make the mental link between the friendly world of (mostly linear) equations that can be solved by formulae and the strange world of those (usually non-linear) that can not.

Figure 14

$$dy/dx=v$$

$$dv/dx=-y$$

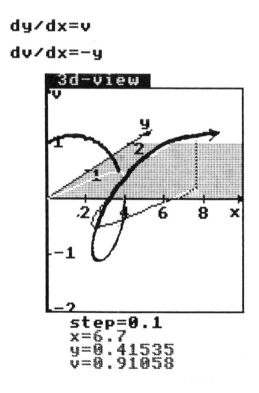

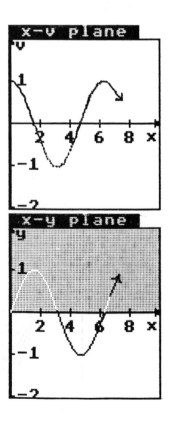

step=0.1
x=6.7
y=0.41535
v=0.91058

IS PROGRAMMING ESSENTIAL ?

We have not explicitly mentioned student programming at all. In the U.K. a body of expertise is growing [Mathematical Association 1985] in which students are expected to handle *short* programs that carry out mathematical algorithms. From here it is intended that they move on to prepared software using the same algorithms in a more powerful interactive manner.

Clearly a spectrum of approaches is possible: some with varying amounts of programming to understand the underlying algorithms, whilst others may exclude programming altogether.

CONCLUDING REMARKS

The diagrams produced in this article are screen-dumps from the BBC computer [T], the Apple Macintosh [H&W], or from [Artigue & Gautheron]. The difference between the graphic configurations and facilities are horrendous. How to exploit the various computer graphics capabilities or to transfer from one to another is a non-trivial problem. But the computer so reduces time and tedium in calculations and one can go so much further with more difficult problems, that the mind is freed to concentrate on the theoretical structure of the mathematics. This has very important implications for how we are to view the curriculum of the future: it is not just a matter of adding an experimental element, but the opportunity to clarify the nature of the mathematical theory itself.

REFERENCES

Artigue M. & Gautheron V. (1983). Systèmes Différentiels: Etude
          Graphique, CEDEC, Paris.
Hubbard J.H. & West B.H. (1985). Differential Equations (to appear).
Sanchez D., Allen R. & Kyner W. (1983). Differential Equations: An
          Introduction. Addison-Wesley.
Takagi T. (1903). A simple example of the continuous function without
          derivative. Proc. Phys.-Math. Japan, 1, 176-177.
Tall D.O. (1982). The blancmange function, continuous everywhere but
          differentiable nowhere. Mathematical Gazette, 66, 11-22.
Tall D.O. (1985). Graphic Calculus for the BBC Computer,
          Glentop Publishers, London.
The Mathematical Association (1985). 132 Short Programs for the
          Mathematics Classroom, Leicester.

*See also supporting papers for the ICMI Conference on "The Effects of
Computers & Informatics on Mathematics & its Teaching" (published as
a separate volume):*
Artigue M., Gautheron V., Isambert I.: L'étude graphique rehabilitée
          par l'ordinateur.
Hubbard J.H. & West B.H.: Computer Graphics revolutionize the teaching
          of differential equations.
Tall D.O.: Visualizing calculus concepts using a computer.

CALCULUS AND THE COMPUTER. THE INTERPLAY OF DISCRETE
NUMERICAL METHODS AND CALCULUS IN THE EDUCATION OF USERS
OF MATHEMATICS: CONSIDERATIONS AND EXPERIENCES

Maria Mascarello
Politecnico di Torino, 10129 Torino, Italia

Bernard Winkelmann
Institut für Didaktik der Mathematik, 4800 Bielefeld, BRD

1 NEW POSSIBILITIES

As is written in many places in this book, the computer is
a mighty mathematical tool, not only for mathematical research, but
even more in the process of applying mathematics, or in the process of
teaching and learning mathematics. In the following, we shall mainly
concentrate on the new possibilities which the computer presents in
the realm of calculus for users and future users of mathematics. By a
user we understand somebody who is interested in mathematics merely
(or mainly) because he uses mathematical models (in particular calculus
models) to solve his (extra-mathematical) problems. Future users of
mathematics are, for example, engineering students, but even those
learning calculus in schools as part of a general education may be
considered under this aspect.

1.1 New possibilities for the user

We first describe the changes in the mathematical knowledge
and habits of the user of mathematics induced by the availability of
sophisticated mathematical software to all who have to rely heavily on
mathematical problem-solving such as engineers, natural scientists,
etc. Whereas the widespread use of such systems (of hardware and software) may to a certain extent seem doubtful, the continuing decrease
of prices and the development of powerful personal computers allows us
to predict that they will be available on microcomputers not only in
research areas, but also in smaller environments by the time, in three
to five years, when our students start their professional careers.

The classic situation of the user of mathematics could have been
described - in a somewhat oversimplified manner - by a huge amount of
passive mathematical knowledge objectivated in monographs, handbooks,
recipes. Normally this knowledge could only be used by being activated
through the active mathematical knowledge of the user himself or by
direct cooperation between the user and a mathematically more
knowledgeable person. In contrast to this, the mathematical knowledge
objectivated in mathematical software can have a far more active
character, e.g. in giving advice and help interactively, offering
possibilities for explorative experiments or answering questions like
a mathematical expert system. But even the more usual numerical software which exists in the form of sophisticated procedures is far more
active than the recipes of the old-fashioned handbooks, since in many

cases these procedures are in fact polyalgorithms. They decide, for
example, with a certain expertise which particular algorithm should be
invoked, depending on the circumstances (cf. Rice 1983, e.g. p.291f).
So the demand for mathematical knowledge on the side of the user has
changed, the emphasis has shifted from a detailed knowledge of the
advantages and disadvantages of specific numerical methods and of the
algorithms themselves to some meta-knowledge of the possibilities of
numerical algorithms in general and their interaction with the concrete
application-situation in these specific circumstances.

As an example let us look at the process of the solution of ordinary
differential equations (cf. Winkelmann 1984). This is indeed an example
of great importance since such equations appear in many fields of
application and are at the heart of applicable elementary calculus. So
if it seems possible to master them at a more elementary level than
hitherto was possible, they could even be regarded as the most
appropriate goal for the teaching of elementary calculus at schools and
colleges. In the education of engineers at technical universities or
similar institutions, where differential equations have always been
part of the calculus sequence, even the beginning calculus could
concentrate more on applications and so give the student a more
realistic and, one hopes, more motivating start.

In the pre-computer age an engineer or scientist who had to handle
differential equations was supposed to have detailed knowledge of
diverse methods for the analytic solution of various elementary types,
to be able to master complicated analytic-algebraic formulas and to
carry out lengthy error-free calculations. How he can use software
which has this knowledge and ability inbuilt, since it can solve more
of the elementary differential equations than a non-specialist mathematician can do (cf. Watanabe 1984). But in building up the model for
the user it is still necessary to fully understand the meaning and
significance of the diverse quantities (variables) and their derivatives
and to be able to relate these to each other in order to set up the
differential equation. And to orderly give it to the computer program,
a thorough intuitive understanding of the mathematical meaning of the
identifiers which appear in the modelling equations is needed, be it
as variables, parameters, initial values, names for (yet unknown)
functions (dependent variables) and so on. If an analytic solution
exists, the program will normally present it as a somewhat confusing
lengthy expression which must be qualitatively interpreted to be understood, namely through looking for simpler special cases, for groups of
specific parameters or initial values, for asymptotic patterns of
behaviour, etc. This process is guided by the intended interpretation
of the solution in the context of the application model. If no analytic
solution exists, the user may give his equation to some ready-made
numerical software. In this case he needs some knowledge to make
reasonable explorative choices of the values of parameters and initial
values; there should be some experiences with numerical phenomena
(pitfalls of computations) and abilities to interpret the numerical and
graphical output of the computer and to use this interpretation

interactively for the new choices of starting points for the next
calculation.

In total, there can be observed a specific shift in the spectrum of
abilities, from precise algorithmic abilities to more complex interpretations, so to speak from calculation to meaning, which in a certain
sense is a reversal of the historical evolution. In this process the
mathematics to be mastered tends to become intellectually more
challenging, but technically simpler.

What does this mean for the mathematical education of the future user?
Of course, there is no direct way from the described activities of the
user to the teaching process; the goal must not be confused with the
way. Understanding and complex interpretations can only be built up by
the personal involvement of the student, by his doing full (but simpler)
examples in all the main steps himself, be it by hand-calculating, by
using interactive symbolic calculators or by programming in some simple
programming languages. This seems necessary in order to get an awareness of the mathematical situations, even if such activities are no
longer part of the final application process. And even if today's
sophisticated mathematical software in most cases need not and can not
be fully understood by the normal user, there must not be totally black
boxes; a principal understanding of simple cases, main ideas or
fundamental restrictions can be gained and seems necessary for proper
use of the now 'grey' boxes.

On the other hand it is quite clear that extensive drill in formal
calculations, fluent structured programming or even perfect handling
of some software package cannot be justified in view of the changed
qualifications needed by the user.

1.2 New possibilities in the teaching-learning process
In the field of teaching methods the computer, if it has
been loaded with the appropriate programs, will function as a de-
technicizing aid, almost as a super hand-held calculator which permits
the pupil to overcome the computational obstacles in the treatment of
more complex problems and more realistic applications, e.g. in dealing
with larger matrices, in the numerical solution of differential
equations, or in the symbolic treatment of more complicated formulas;
this will serve to widen the potential scope of mathematics education
in terms of content. On the other hand, a computer equipped with the
appropriate languages and environments can become an instrument for
solving problems in the hands of the student (interactive programming);
in this case, the student tends to understand techniques more on the
cognitive level, and no longer mainly on the level of skill. Beyond
that, the computer, with its possibilities for illustration and symbol-
ization, will provide opportunities for providing more comprehensive
and rapid mathematical experience.

This presents problems and tasks for educators mainly on two levels.
On a more technical level, there is the necessity to provide more

suitable software. On a more fundamental level, the problem is to achieve a balance in the quantitative and qualitative relation of new and old goals and methods as well as to determine trends.

The computer creates new opportunities for analysis instruction, e.g.
- numerical and graphical illustrations,
- more complex and more realistic applications,
- a language in which to describe traditional calculus,
- CAL (computer-aided learning) in its various forms.

Some traditional motivations for treating conceptually exacting analysis in school can, however, no longer be maintained without further discussion, for instance:
- calculations such as finding extreme values or areas can be easily programmed without analysis,
- practical applications, as in physics or technology, use discrete methods in computer programs.

This results in a crisis: the legitimacy of traditional analysis in school is challenged; educators will have to make clear to the general public, and the teacher will have to explain to his pupils asking critical questions, how and why treatment of continuous analysis still makes sense nowadays.

In Section 3 we shall report on some experiments concerning the use of informatic tools in teaching basic mathematical courses at the Politecnico of Turin, Faculty of Engineering Sciences. We emphasize that the choice here has been to keep the teaching of calculus traditional, giving in the main course of the lectures some basic informatic notions and devoting special exercise-sections to "calculus at the computer".

2   THE INTERPLAY DISCRETE - CONTINUOUS
2.1   General considerations
Although the role of applications, specifically those of analysis, has been changed by both the growing number of disciplines using corresponding models and new methods, particularly the use of computers, an understanding of fundamental approaches in which mathematizations take place remains indispensable. Examples of such concepts are:
- variable quantity, change
- functional connection
- local rate of change
- average value
- cumulation.

We shall refrain from discussing here how far traditional mathematics education was able to attain the goal of teaching these.

Now it is evident that these central approaches to mathematical applications can be implemented both by discrete and by continuous

conceptualizations. Corresponding to such continuous concepts as
function, differential equation, derivation, weighted integral and
integral, are the corresponding conceptualizations in discrete analysis,
namely, sequence and time series, difference equation, difference,
arithmetical mean value and sum.

These discrete concepts are obviously technically and intellectually
much simpler than their continuous counterparts.

In the following we will give some justifications which are in our
opinion crucial in answering the inevitable question now raised: "Why
use the concepts of continuous analysis in teaching at all?"

(a) Insufficiency of continuous analysis for obtaining
     concrete numerical results
     Let us recall some of the facts; most integrations cannot
be executed analytically, but only numerically; this is all the more
true for solving differential equations. But even tasks as simple as
determining the extremes of a familiar function like x * sin x will
require numerical methods. School mathematics has hitherto confined
itself in a rather unnatural way to problems involving classes of
functions which were solvable by analytic methods. It has paid dearly
for this with heavy losses in reality, content and relevance. The
analogue is true for classical university courses in, for example,
elementary differential equations.

(b) Most concrete models of analysis have a discrete basis
     This is first evident in the social sciences or in popula-
tion biology, when the quantities to be modelled are numbers of items
or individuals, or monetary units, which cannot be subdivided at will.
But in physics, too, for instance, most models start discretely: even
disregarding the fact that the universe is finite in principle and
structured in particles, and that there are quanta (i.e. smallest
units), it is a fact in the case of quantities which are usually
conceived of as being continuous, and mathematized accordingly, that
concrete models based say, on results of measurements, will start
discretely simply because continuous functions cannot be obtained as
results of series of measurements which yield only discrete sequences
or time series (this does not hold, of course, for modellings based on
theoretical approaches).

(c) The transition from models to concrete numerical results
     in general, cannot be accomplished without continuous
     analysis
     This is true, for one thing, because of the rounding errors
which inevitably occur in numerical computing, and have to be controlled
by a superordinate model. A second, deeper reason follows from a closer
look at the discrete aspects mentioned in points (a) and (b): it is
found that the step-widths used in (a) and (b) are basically independent
of each other. The density of the values measured in the measuring
process is generally determined according to practical aspects. It

results from consideration of information content and "cost". One of
the most fundamental hypotheses for determining the step-widths is that
a diminution of step-widths may yield more exact results, but basically
none which differ in principle. The phenomena which are to be observed
and/or described are considered to be invariant with respect to the
step-width used in the observations, provided it is sufficiently small.
This fits in with the assumption that the corresponding limits exist.
It is only on the basis of this assumption that the measuring process
can be carried out in a discrete way chosen by practical considerations. In this case, however, the phenomena concerned are basically
invariant with respect to the step-width, and are thus best described
in mathematical models which do not explicitly contain step-width. The
fact that the step-width, with which the measuring data were obtained,
is only of marginal importance even for the model, explains why the
step-widths used, say, to solve numerically the corresponding differential equations, will generally be completely independent of the step-
width used in measurement. The latter are determined by practical
criteria such as cost and the precision required.

This fundamental consideration, which is decisive in what follows, has
been formulated here only for the special case where the results of
discrete measurement are used as a starting point. It is true, in an
analogous way, for all the other cases in which mathematizing and
modelling is done by analysis. In particular, this consideration helps
us to explain why some disciplines in which the natural structure is
discrete, such as number of individuals in population biology, nevertheless use continuous models, despite the fact that this would seem
inappropriate at first glance: the impact of such small changes on the
phenomena concerned in the respective models is only marginal.

This behaviour is of course not valid for all mathematical models in
the mentioned sciences or other domains. But it is in a sense typical
for calculus models: if this behaviour is not observed in a specific
situation, then normally we should use really discrete models, and if
we – for technical reasons – nevertheless use some calculus models, we
should be aware of our improper use and of possible difficulties in
interpreting results. This may happen for example if we try to
consider "fractal" phenomena in nature, such as natural border lines
(of islands, leaves of trees, etc.). Here indeed the application, say,
of formulas for the length of a curve does not make much sense.

2.2 The context of dynamical systems
Dynamical systems (systems of time-independent explicit
first order ordinary differential equations) appear as rather natural
mathematical models for many situations in a variety of disciplines
such as the physical, biological or economic sciences. Here typically
we have to distinguish between situations where a natural step-width
exists whose value influences the phenomena, and situations in which
this is not the case. In both cases, modelling with (discrete)
difference equations is possible and adequate; but whereas in the
former case, the step-width of the difference equation has to be equal

to that of the underlying situation, in the latter it may be chosen as
a free parameter which suggests that the use of differential equations
might be more natural.

As an example, consider the logistic growth of a (biological)
population. If the generations of the population are distinct, as with
certain bugs, there may be observed oscillations and fluctuations of
the population, which are easily modelled and explained in the context
of the difference equation, but would disappear in the transition to
the corresponding differential equation (if it were not explicitly
modelled by including a time lag which would induce the same fluctuations but would exclude the resulting equation from what is normally
considered a differential equation in mathematics). But if generations
are not distinct and population oscillations are slow compared to
normal reproduction times, modelling with (logistic) differential
equations seems adequate, even if there were always only discrete
points in time where new offspring could be noticed.

3 EXPERIENCES

In this section we would like to report on some experiments
concerning the use of informatic tools in teaching basic mathematical
courses at the Politecnico of Turin (Italy), Faculty of Engineering
Sciences. These experiments refer in particular to the courses
Mathematical Analysis 1° and Mathematical Analysis 2° given to students
of Mechanical Engineering in the years 1980 to 1983, using pocket
computers. This activity was continued in 1984 and 1985, in the same
courses, using such micro computers as the Sharp MZ803 and IBM PC. At
this second stage, the experiment has concerned a restricted number of
students, selected on the basis of a test.

While we refer to Mascarello-Scarafiotti (1985) and to Boieri et al.
(1984) for the general aims, the list of the themes and the obtained
results of this experience, we should like to detail here some topics
and contents, and add some final comments, as a 'proof' of what we
asserted in Section 2.

Let us begin by observing that, to carry out the experience in a correct
way, it has been necessary, of course, to rely on basic informatic
arguments. To this end, in the main course of the lectures, the
teacher, after giving some notions of the theory of formal languages,
then introduced machine-numbers and algorithms for floating-point
arithmetic computations. Always in this direction and already in the
first part of the course, some proofs of classical analysis results
were presented in a computational form.

One of the most important experiments concerned the study of the
dynamical system using microcomputers. More specifically, we began in
Mathematical Analysis 1° with the study of discrete dynamical systems,
which was introduced after the study of sequences defined by recurrence
formulae. As a natural continuation, in Mathematical Analysis 2° we
considered continuous dynamical systems, giving a formal expression of

the qualitative results. Finally we return to the use of microcomputers
to find numerical results; this was done in order to check the known
results of the theory, and also to conjecture new results, concerning
open problems. To be more definite, we begin by briefly listing the
contents of the exercise-sections concerning dynamical systems
(Mathematical Analysis 2°):
- Cauchy problem for first order ordinary differential equations;
  solutions at the microcomputer, comparing the methods of Euler and
  Runge-Kutta.
- First order systems of ordinary differential equations, and in
  particular autonomous systems; visualization of the trajectories in
  the phase plane.
- Second order ordinary differential equations: solutions on the microcomputer of some non-linear equations of particular significance in
  applications, such as the pendulum and the Duffing equations.
- A numerical approach and simulation on the microcomputer of the
  trajectories for some problems which are still open in their
  qualitative aspects, as for example the mathematical model of the
  Lorenz attractor.

Let us detail further the content of some exercise-sections, which
appear to us particularly significant from the didactic point of view.

i) The student, knowing the classical analytic solutions of linear
equations with constant coefficients, and having some basic notions of
the stability theory, is invited to 'solve' the equation $\ddot{x} + k\dot{x} + x = 0$
on the microcomputer and to visualize the trajectories in the phase
plane (without any direct assistance from the teacher). Figure 1 shows
some drawings obtained by a student.

A discussion with the students followed concerning the validity of the
results obtained in this way; particularly surprising is the second
picture, where closed trajectories appear for $k \neq 0$.

ii) The students "solve" on the microcomputer the pendulum equation
$\ddot{x} + \sin x = 0$ by the Runge-Kutta method. Figure 2 shows the drawing
obtained by a student.

We can observe that the picture seems satisfactory from a numerical
point of view. Some qualitative aspects of the solutions are under-
lined by the teacher, as a check of the known results from the theory.

iii) The student is invited to simulate on the screen the trajectories
of the equation of the Lorenz attractor:

$$dx/dt = -sx + sy$$

$$dy/dt = rx - y - xz \quad \text{with} \quad s = 10, \ r = 28, \ b = 8/3$$

$$dz/dt = -bz + xy \ .$$

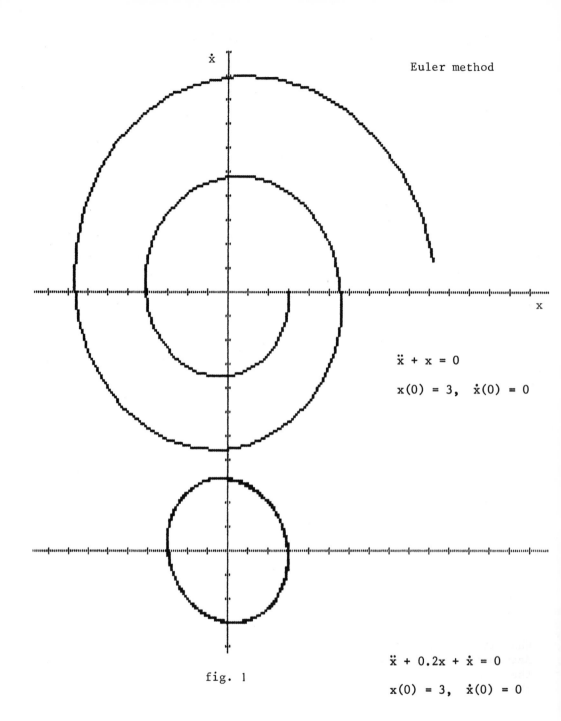

Euler method

$$\ddot{x} + x = 0$$

$$x(0) = 3, \quad \dot{x}(0) = 0$$

fig. 1

$$\ddot{x} + 0.2x + \dot{x} = 0$$

$$x(0) = 3, \quad \dot{x}(0) = 0$$

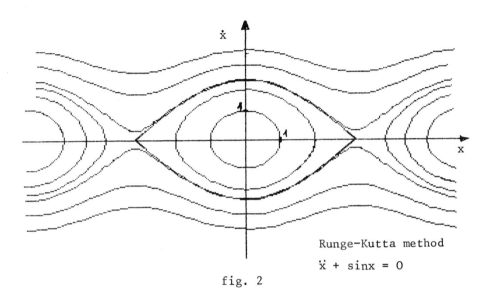

Runge-Kutta method

$$\ddot{x} + \sin x = 0$$

fig. 2

In fig. 3 there is a picture obtained by a student (the completion of the program required a certain informatic ability, due to the complications arising from the 3-dimensional representation of the trajectories in the (x, y, z)-space).

No comparison was attempted with known qualitative results since the existent literature on the subject seems to be too far advanced for a second year engineering student. However, a comparison was possible with what might be expected from the physical phenomenon (such as fluid turbulence phenomena).

What it is very important to emphasise, is that at this stage (end of Mathematical Analysis 2°) the student was able to evaluate correctly the results obtained from the computer, namely, to take into account the discrepancies which may occur between numerical solutions and analytic solutions, keeping in mind that his final objective is the interpretation of the physical phenomenon.

Our considerations have shown that even today where discrete working computers are used for handling calculus models (so far as applications

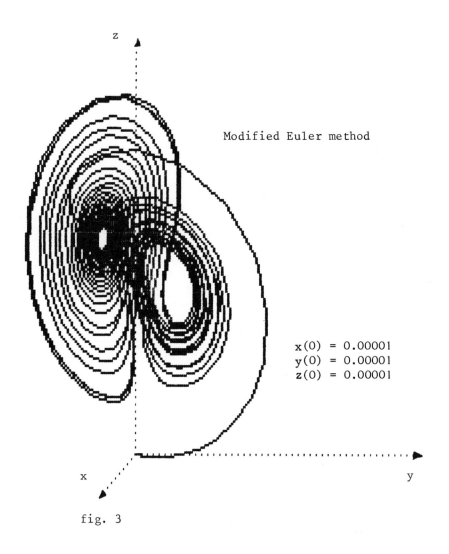

fig. 3

are concerned), continuous analysis cannot be dispensed with when
describing problems for which analysis has been classically used.
This, however, need not lead to the conclusion that analysis education
at school or universities should go on as before. The discussion has
shown the function of continuous analysis in applications, and teaching
must be done in such a way that this function is fulfilled. This
requires that the transition from the discrete to the continuous model
be experienced by the students and that the respective particular
possibilities and limitations of the model type in question be

perceived. To us, it would seem dishonest to try to explain to the
student the importance of analysis for applications by means of unrealistic and oversimplified minimum-maximum tasks. Rather, it would
seem crucial to have the student at least begin to assess the usefulness of the various components of the system of analysis, i.e. concepts,
approaches, calculi, translation schemes in practical applications.
This goal should be attained by appropriate problem solving in the
classroom; and explication should play a subordinate part. It remains
to be seen how a balance between the individual components can be
achieved. The following aspects, however, should be included in any
case:

a) Analysis teaching should include treatment and study of discrete
models. This leads to numerical computations. It does not necessarily
imply explicit teaching of numerical mathematics, but requires
including important numerical basic facts such as propagation of errors.

b) Establishing models is an important activity which must not be
neglected in favour of interpreting models. In particular, this means
that the techniques of finding suitable functions are as important as
discussing functions.

c) The role and function of (continuous) calculus must be developed in
an appropriate way. It cannot be used to obtain numerical results,
save in exceptional cases: it can, however, guide and direct the use of
numerical methods.

d) The recent development of computer science has established
techniques of description, in particular programming languages, which
permit the precise description even of complicated processes such as,
for instance, the algorithms necessary for symbolic differentiation.
Mathematics teaching should increasingly make use of this.

REFERENCES

Boieri, P. et al. (1984). 'Personal computers in teaching basic
          mathematical courses'. SEFI Annual Conference: The Impact
          of Information Technology on Engineering Education.
          Erlangen.
Mascarello, M. and Scarafiotti, A.R. (1985). 'Computers experiments on
          mathematical analysis teaching at the Politecnico of Torino
          (Italy)'. ICMI Congress: The influence of computers and
          informatics on mathematics and its teaching. Strasbourg 2°
          Document du Travail. Rapporto interno n. 5. Politecnico
          di Torino. Dipartimento di Matematica.
Rice, J.R. (1983). Numerical Methods, Software, and Analysis. IMSL
          Reference Edition. New York: McGraw Hill.
Watanabe, S. (1984). 'An experiment toward a general quadrature for
          second order linear ordinary differential equations by
          symbolic computation'. J. Fich Ed. EUROSAM 84. Interna-
          tional Symposium on Symbolic and Algebraic Computation.

Cambridge, England: July 9 - 11, 1984. Berlin: Springer, 13 - 22.

Winkelmann, B. (1984). 'The Impact of the Computer on the Teaching of Analysis'. Int. J. Math. Educ. Sci. Techn. 15, 675 - 682.

COMPUTER-BASED SYMBOLIC MATHEMATICS FOR DISCOVERY

K.D. Lane, A. Ollongren and D.R. Stoutemyer

K.D. Lane, Harvey Mudd College, Claremont, CA 91711, USA.
A. Ollongren, Leiden University, 2300 RA Leiden,
                                        The Netherlands.
D.R. Stoutemyer, University of Hawaii, Honolulu,
                                        Hawaii 96822, USA.

INTRODUCTION

In the present contribution three authors from different
schools and backgrounds present some views on the influence of
computer-based systems for carrying out symbolic mathematical operations.

We consider the influence of such systems on the mathematics curriculum
and give elementary as well as advanced examples illustrating the process
of discovery. In Section 3 suggestions for further discovery are
collected.

1 A Curricular Project

Events and developments of the past several years have
brought pressure for reform and change in the undergraduate mathematics
curriculum. The evolution of computer science and related equipment,
the mathematization of subject areas outside of the physical sciences,
and the gradual redefinition of student skills and preparedness are
important component forces in the production of these pressures. In
June of 1983 the Colby College Mathematics Department was awarded a
grant by the Alfred P. Sloan Foundation "for the development of a new
curriculum for the first two years of undergraduate mathematics in
which discrete mathematics will play a role of equal importance to that
of calculus" [Hosack et al, 1983]. An important aspect of this
curriculum was that it was to reflect the existence and widespread
availability of computing machinery and computational systems.
Computer based experiences were to be provided to motivate and illustrate concepts whenever appropriate. Throughout the proposed
curriculum, the existence of new technologies and its effect on the
relative importance of subject matter was to be considered.

During the development of the proposal it became obvious to those
involved that the calculus curriculum would have to be substantially
revised. Equal time for discrete mathematics seems to imply that the
standard three semester single and multivariable calculus sequence
would need to be distilled into a single one year course. Not only
would the syllabus for such course be different, but the spirit of the
course needed to be radically altered. A textbook [Small et al, 1983]
was written and the course was first taught in the fall of 1983. Two
aspects of the course were somewhat novel: (1) Single and multivariable

topics were done concurrently, and (2) Symbolic manipulators were incorporated.

The incentive for a fresh approach to calculus comes not only from the pressure of introducing discrete mathematics but also from a sense of dissatisfaction with the lack of success of the traditional course in teaching processes and methods of mathematical thinking. Students often feel that the real meat of the course is the computation of derivatives, integrals, power series, or any of the other manipulative activities. Attempts to get students to focus on analysis and synthesis often end in failure. If the instructor focuses on the ideas of calculus, he is viewed as being hopelessly and inappropriately stuck on useless abstractions or irrelevant garnish.

It is no small wonder that many of our students have these attitudes. Consider the daily exercises and examination questions that students ultimately use to gauge the relative importance of topics in the course. Student perception of the importance of computational activities is reinforced by the fact that these assignments are often difficult and take a lot of time to complete.

Central to the Colby calculus course are three "Fundamental Processes" of calculus. As topics were woven together to design the course, the prime consideration was the reinforcement of these fundamental issues. Somewhat arbitrarily the Fundamental Processes were identified as:

(1) approximation;
(2) transformation;
(3) comparison.

In order to be able to spend more time in the classroom developing ideas, the Colby group decided to introduce the symbolic system MACSYMA in the experimental calculus course. The hope was that time normally spent on developing computational skills, such as techniques of integration, would be greatly reduced if this technology was made available to students. MACSYMA was an optional tool in a section of the calculus course taught during the 1983-84 academic year.

During the fall 1984 term, students in the experimental calculus course were required to use the symbolic system MAPLE. Students were given instruction on the routine use of the system, including the protocol for invoking the system, for editing expressions and for accessing on-line or printed documentation. This did not include the use of "programming" control constructs such as procedures, loops, and conditionals. Most systems have a rich set of commands that are directly executable in a straightforward "calculator mode", and many explorations and projects require no more than a modest subset of these commands together perhaps with the use of assignment to preserve results for use in subsequent expressions. For example:

$$\text{p: DIF } (-93 \text{ x}^4 \text{ y}^3 + 439/2 \text{ x}^2 \text{ y}^2 - \text{x y}^5, \text{ x});$$

might assign the partial derivative $-372\ x^3\ y^3 + 439\ x\ y^2 - y^5$ to the
variable p, after which the command

$$FACTOR\ (p + 163308\ x^4);$$

might produce the equivalent form $(372\ x^3 + y^2)(439\ x - y^3)$.
Proficiency in such elementary use can be promoted by straightforward
exercises such as

"Use the computer algebra system to factor $x^{16} - 64\ y^{24}$."

2 Some Specific Examples

In the beginning we had only some vague notion of how we
might utilize computer algebra in the calculus course. Our plan was to
proceed cautiously by experimentation and to see what might develop!
We have been genuinely surprised by much of what occurred and are
excited about the prospects for yet more of the same. We now consider
some examples to illustrate the possibilities.

Among the more interesting applications of derivative concepts is curve
sketching. Although this activity ought to be perfect for illustrating
and reinforcing newly learned concepts, there is a tendency no longer
to include it in calculus courses. After all, typically students have
trouble learning the overall structure of the process, since attempts
to utilize the calculus tools usually abort when some simple computational mistake is made. What is hoped to be an illustration of the
power of calculus ends up being an exercise in drudgery. Without
having enough successful complete experiences with these techniques,
the overall picture is not grasped by the student. The intending goal
in teaching the sketching techniques is lost in a sea of confusion over
manipulation.

In the presence of a symbolic manipulator, these exercises take on an
entirely different character. The introduction of the symbolic system
elevates the level of sophistication of this particular exercise.
Consider what happens with a specific example.

Exercise: Sketch the graph of $f(x) = \dfrac{x^2-4}{x^2-1}$. Indicate all "interesting"
features.

In what follows we illustrate a possible student session using an
experimental version of the muMATH^{tm} system. This version is scheduled
for distribution some time during 1985 for the IBM-PC and other
computers using the similar MS-DOS operating system (details from The
Soft Warehouse, Honolulu, Hawaii, 96822, USA). The example is well
within the capabilities of virtually all systems, some of which are
referenced in (Stoutemyer, to appear).

The system used here prompts the user with a numbered label beginning
with the letter "i" for "input" and followed by a colon. The user then
enters an expression terminated by a semicolon. The system then

displays the computing time in seconds if it is nonnegligible compared
to the computer clock resolution. Next, the system displays a numbered
label beginning with the letter "o" for "output", followed by a colon
then the corresponding result. The outputs can be numbers, expressions
or function plots.

Previous inputs and outputs can be recalled for editing or for use in
subsequent expressions. For ease of typing, inputs use "/" for
division and "^" to denote raising to a power. For ease of reading,
outputs use raised exponents and use built-up fractions where it is
attractive to do so. The entire dialogue can automatically be recorded
on diskette for subsequent editing, printing or reentry. The students
would be familiar with such details from their earlier trivial
exercises.

First, we illustrate how the student can use a sequence of built-in
functions to accomplish the composite task of the above exercise.

2.1 Curve Sketching

i1:   (x^2-4)/(x^2-1);

o1: $\dfrac{x^2-4}{x^2-1}$.

We can now compute the zeros of the expression:

i2:   solve(o1=0,x);

o2:   [x=-2, x=2] .

Note that the student can refer to the equation in question as "o1" .
We can also compute the singularities of the expression. The next
command extracts the denominator:

i3:   denominator (o1);

o3: x^2-1

i4:   solve(o3=0,x);

o4:   [x=-1, x=1] .

This is a calculus course – compute a derivative:

i5:   diff(o1,x);

o5: $\dfrac{6x}{(x^2-1)^2}$.

The task for the student is now focused on what particular question must be asked of the mathematical object at hand: What do you do with the first derivative? The student may choose to compute the zeros and singularities of this derivative:

i6:   solve(o5=0,x);   i7:   solve(denominator(o5)=0,x);

o6:   [x=0]   o7:   [x=-1, x=1] .

The process continues in this fashion. The student must consider the information at hand and decide how to further process it. The student is forced to consider the relevant questions. The computation is discounted. Suddenly again we are interested in teaching curve sketching!

Now we will illustrate a more open-ended problem that entails more discovery.

We would give the students time to ponder the following mock project assignment for several minutes before commencing the demonstration:

Project: Use your computer algebra system to explore inter-relationships among the coefficients of $(x+y)^n$, expanded for increasing n. Discuss the issues listed below and any other relevant ones that you discover:

    a)   the number of terms;

    b)   relations among the exponents in successive terms;

    c)   symmetries among the coefficients for a particular n;

    d)   relations among coefficients for two successive values of n;

    e)   relations between a coefficient and factorials involving the corresponding exponents;

    f)   the asymptotic growth of the largest coefficient with n.

    g)   the asymptotic growth in computation time with n.

Include plots that helped lead to your discoveries or that vividly summarize them. Include proofs if you can. Do not worry if you cannot decisively address all of these issues.

Superficially, this particular example would seem most appropriate at the point in the curriculum just before first exposure to binomial expansion. However, some parts of the project would require more maturity. It certainly does not ruin such a project if the students already know some of the answers. Such reinforcement can be beneficial. Moreover, elementary demonstration examples permit the students to concentrate on the exploratory techniques without being distracted by a flood of new mathematical facts.

We have enclosed the spoken narration below in quotes to help distinguish it from the computer dialogue with which it is interspersed.

2.2.1 Coefficient Patterns

"Well class, here is how I might proceed with this project if it were as new to me as it is to you. First, I would try a few successive values of n to see what that reveals:"

i1: EXPAND: TRUE;   "Let's set the expansion control variable to
o1: TRUE   request automatic expansion until further
                     notice."

i2: (x+y)^0;
o2: 1   "I knew this result, but such degenerate cases
                     may be an important part of a pattern."

i3: (x+y)^1;
o3: x + y   "This is the only other degenerate case that I
                     can perceive."

i4: (x+y)^2;
o4: $x^2 + 2xy + y^2$

i5: (x+y)^3;
o5: $x^3 + 3x^2y + 3xy^2 + y^3$

i6: (x+y)^4;
o6: $x^4 + 4x^3y + 6x^2y^2 + 4xy^3 + y^4$

i7: (x+y)^5;
o7: $x^5 + 5x^4y + 10x^3y^2 + 10x^2y^3 + 5xy^4 + y^5$

"It appears that there are n + 1 terms when $(x+y)^n$ is expanded."

"The exponents of x appear to start at n and decrease by 1 to 0 in each successive term while the exponents of y appear to start at 0 and increase by 1 to n in each successive term."

"The coefficients appear to be symmetric about the centre term or central pair of terms."

"The end coefficients appear always to be 1."

"The penultimate coefficients appear always to be n."

"I can't yet see how the other coefficients relate to n."

"However, the project assignment first suggested looking for relations between the coefficients for successive values of n, and I'm not too proud to accept a hint."

"It does appear that the coefficient 6 in o6 equals the sum of the coefficient 3 directly above and the coefficient 3 to its left in o5. In fact, this "sum of above and to its left" pattern holds for every coefficient if we imagine coefficients of zero surrounding the displayed nonzero coefficients! This remarkable pattern seems too simple to be true. I'll check n = 6 to see if it provides a counterexample:"

i8: (x+y)^6;
08: $x^6 + 6 x^5 y + 15 x^4 y^2 + 20 x^3 y^3 + 15 x^2 y^4 + 6 x y^5 + y^6$

"The pattern still holds!"

"How far should I explore? I could write a procedure with a loop that increments n by 1 each time and compares the coefficients in $(x+y)^n$ with the appropriate sums of those in $(x+y)^{n-1}$ until a counterexample is encountered or until the computer runs out of memory. I can run the program overnight. Even if the program does not find a counterexample by tomorrow morning, the increased evidence for the rule would encourage me to seek a proof. Parts f and g of the project may even permit me to estimate how large n can become before I run out of memory space or patience. However, I'll postpone writing, debugging and starting that program until I have no further ideas for quick interactive experiments."

"The next part of the assignment is to discover a relationship between each coefficient and factorials involving the corresponding exponents. Well, 0!=1!=1, 2!=2, 3!=6, 4!=24, 5!=120 and 6!=720; so the coefficient of $x^k y^{n-k}$ is clearly not simply n!, k! or (n-k)! Thus the coefficient must be some composition of factorials if it involves factorials at all. Moreover, since the coefficients are symmetric, the composition should be symmetric in k and n-k."

"I cannot yet perceive an obvious relation, so I will give up on that - at least for a while. Perhaps an inspiration will occur after some experience with other aspects of the project or after a sufficient incubation period."

2.2.2 Coefficient Growth

"The next suggestion is to study the asymptotic growth in the largest coefficient as n increases. The largest coefficient appears to always be the central one when n is even or either of the equal central pair when n is odd. Through n = 6 the growth is rather modest, so rather than continuing to creep along by uniform increments of 1, let's next try n = 8, 16, 32,..., doubling n each time until we run out of memory or patience."

"When n is even, the center coefficient is that of $x^{n/2} y^{n/2}$. Accordingly, we can use the built-in coefficient extraction function as follows to avoid cluttering our screen with superfluous information:"

i16: COEF ((x+y)^8, x^4 y^4);
0.3 sec.
o16: 70

i17: COEF ((x+y)^16, x^8 y^8);
0.8 sec.
o17: 12870

i18: COEF ((x+y)^32, x^16 y^16);
2.6 sec.
o18: 6010 80390

i19: COEF ((x+y)^64, x^32 y^32);
7.6 sec.
o19: 1832 62414 09425 90534

"This sequence will soon become too time consuming for interactive
exploration. If I decide to do more, I'll write a procedure containing
a loop and run it overnight. A vague pattern of sorts has already
emerged anyway."

"Considering also the previously done cases n = 1, 2 and 4, each
doubling of n appears to approximately double the number of digits.
Thus, the number of digits in the largest coefficient appears to be
roughly proportional to n."

"Since the number of digits in a coefficient is approximately proportional to the logarithm of the coefficient, the coefficient itself
appears to grow approximately exponentially with n. Logarithms to
differing bases are proportional, so the choice of base is not crucial.
However, since we are interested in the number of decimal digits, let's
plot the piecewise linear interpolant of LOG_{10} (largest coefficient) as
a function of n ≥ 1 to see how well it approaches a straight line with
increasing n:"

i20: LINEARSPLINE ([1,LOG(1,10)], [2,LOG(2,10)], [3,LOG(3,10)],
     [4,LOG(6,10)], [5,LOG(10,10)], [6,LOG(20,10)], [8,LOG(o16,10)],
     [16,LOG(o17,10)], [32,LOG(o18,10)], [64,LOG(o19,10)]);

o20: lower left corner = (1, 0), upper right = (64, 18.26)

See Figure 1.

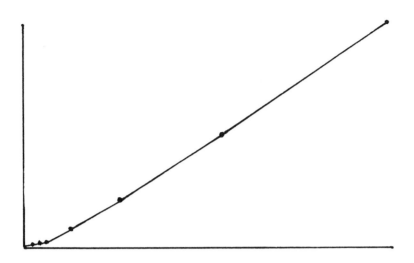

Figure 1.

"The semi-log plot appears to approach a linear asymptote quite well, so let's fit a line through the two largest measurements to use for predicting the number of digits with larger n:"

i21: SOLVE(s-LOG(o18,10))/(LOG(o19,10)-LOG(o18,10))=(n-32)/(64-32), s);
o21: {s=0.296379 n -0.705209}.

"$(x+y)^n$ has n + 1 coefficients varying from 1 through this maximum number of digits s. Their average appears to be more than half s, so let's conservatively estimate the total space as n*s:"

i22: n RHS(o21 [1]);
o22: 0.296379 n^2 - 0.705209 n .

"Thus, n = 128 would use a total number of digits about:"

i23: SUBST(o22, n: 128);
o23: 4765.61.

"My computer has enough memory for simultaneously holding a few tens of thousands of digits total. Consequently, allowing a generous margin for other numbers created during the expansion, there should be sufficient room for one or two more doublings."

2.2.3 Computing Time

"Now let's estimate how much time these larger values of n
will require: The computing time appears to increase by a constant
factor of about 3 as n increases by a factor of 2. This suggests an
asymptotic power-law dependence: $t = c\ n^p$. Just as exponential growth
is associated with a straight-line semi-log plot, power-law growth is
associated with a straight-line log-log plot. The choice of base is
not crucial, so I'll use the natural log:"

i24: LINEARSPLINE([LN 8, LN 0.3], [LN 16, LN 0.8], [LN 32, LN 2.6],
     [LN 64, LN 7.6]);
o25: lower left corner = (2.08, -1.204), upper right = (5.16, 2.03).

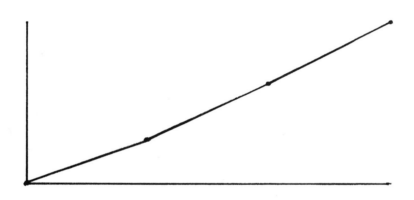

Figure 2.

"The log-log plot [shown in Figure 2] appears to approach a linear
asymptote quite well, so to fit a line through the logarithms of the
two largest measurements to use for prediction:"

i25: SOLVE
     ((LN t - LN 2.6)/(LN 7.6 - LN 2.6) = (LN n - LN 32)/(LN 64 - LN 32),t);
o25: $\{t = 0.0191558\ n^{1.54749}\}$.

"Thus, I guess that if we don't run out of space, the number of
seconds required to compute $(x+y)^{256}$ would be about:"

i26: RHS(o25 [1]);
o26: $0.0191558\ n^{1.54749}$.

i27: SUBST(o26, n: 256);
o26: 64.9389.

"This is feasible to try right here in class, but my plan was to
compute and compare expansions for all successive n up through the
maximum allowable by the memory. Consequently, our total time as a
function of the last value of n, which I'll call m, would be at least:"

i28: SUM(o26, n, 0, m);

$$o28: 0.0191558 \sum_{n=0}^{m} n^{1.54749}.$$

"The system was unable to find a closed form for this indefinite sum,
and I wouldn't be surprised if none exists in terms of the elementary
functions with which we are all familiar. Consequently, let's try
approximating the sum by an analogous integral:"

i29: DEFINT(o26, n, 0, m);

$$o29: 0.00751948 \, m^{2.54749}.$$

"Now we can estimate how far we can get in a 12-hour computation:"

i30: SOLVE(o29 = 12*60*60, m);
o30: {m = 450.107}.

"It appears that an overnight run will indeed be the right order of
magnitude for proceeding by increments of 1 until we exhaust the memory
available for numbers." ...

2.3 Other Considerations

The space limitation here prevent us from completing the
scenario. However, the demonstration would continue on to the point of
showing how computer algebra can be used to support theorem proving.
Next we would distribute a sample written report based on the demonstrated experiments and proofs. Then we would distribute an appropriate project assignment of this nature for the students to do.
Section 3 contains a list of such projects addressing a variety of
mathematical topics. This list is the beginning of one that we plan to
collect and refine for publication. Suggestions and additions will be
gratefully acknowledged.

It might be wise to give each student two or three choices, because
their individual insight could vary erratically on open-ended problems
such as these. For mathematical topics that suggest a great many
projects, it might be especially motivating to allocate the choices in
such a way as to collectively attack most of the problems, with each
report then presented to the group so as to pool experiences.

3   Some Discovery Projects Using Computer Algebra

3.1   Elementary Algebra:

1.   Experiment with your computer algebra system to form a conjecture
     about how the reduced form of the algebraic expression
     $(x^m - 1)/(x^n - 1)$ depends on m and n.   Then, use the system to help
     prove your conjecture inductively.   Discuss the growth of comput-
     ing time (and perhaps also space) with m and n.

2.   Use your symbolic math system to factor $x^n \pm y^n$ over the integers
     for increasing n.   Form some conjectures about the number and form
     of the factors versus n.   For example, are the factors of n rele-
     vant?   How do the coefficient magnitudes vary with n?   Try proving
     your conjectures.   What is the asymptotic growth of computing time
     (and perhaps also space) with n?

3.   Using computer algebra, determine the reduced forms of

     $1/(1 - x^2/(3 - x^2/5))$,

     $1/(1 - x^2/(3 - x^2/(5 - x^2/7)))$,

     $1/(1 - x^2/(3 - x^2/(5 - x^2/(7 - x^2/9))))$,

     etc., with the constants being successive odd integers.   Super-
     impose plots of these functions.   Does the sequence of functions
     appear to be converging to a well known function?   Is the con-
     vergence monotonic?   How does the computing time appear to grow
     asymptotically with the number of operations in the truncated
     approximation?   If you know a power-series approximation for the
     same function, how does it compare in speed versus accuracy for
     different ranges of x?

3.2   Matrices and Determinants:

1.   Use your computer algebra system to form the matrix products

     $$\begin{bmatrix} a & 1 \\ 1 & 0 \end{bmatrix} \begin{bmatrix} b & 1 \\ 1 & 0 \end{bmatrix}, \quad \begin{bmatrix} a & 1 \\ 1 & 0 \end{bmatrix} \begin{bmatrix} b & 1 \\ 1 & 0 \end{bmatrix} \begin{bmatrix} c & 1 \\ 1 & 0 \end{bmatrix}, \quad \text{etc.,}$$

     each time including one more matrix until you can infer the
     general form of the elements in the product.   Then, see if you
     can use the system to help inductively prove your general form.

2.   For each of the following exercises, use your computer algebra
     system to compute successively higher-order determinants of the
     indicated family until you can conjecture the general form.   Try
     to prove your conjecture.   What is the nature of the growth in
     computing time and space versus order?   Beware that the behaviour
     may differ for odd and even orders.   Also, you may need to expand,

factor, or otherwise rearrange the nominal results in order to reveal the most regular form.

a)
$$\begin{vmatrix} 1 & 1 & 1 & 1 \\ 1 & a+1 & 1 & 1 \\ 1 & 1 & b+1 & 1 \\ 1 & 1 & 1 & c+1 \end{vmatrix}$$

d)
$$\begin{vmatrix} 1 & -1 & 0 & 0 & 0 \\ x & h & -1 & 0 & 0 \\ x^2 & hx & h & -1 & 0 \\ x^3 & hx^2 & hx & h & -1 \\ x^4 & hx^3 & hx^2 & hx & h \end{vmatrix}$$

b)
$$\begin{vmatrix} 1 & a & b & c \\ a & 1 & 0 & 0 \\ b & 0 & 1 & 0 \\ c & 0 & 0 & 1 \end{vmatrix}$$

e)
$$\begin{vmatrix} 1 & a & a^2 & a^3 \\ 1 & b & b^2 & b^3 \\ 1 & c & c^2 & c^3 \\ 1 & d & d^2 & d^3 \end{vmatrix}$$

c)
$$\begin{vmatrix} 0 & 1 & 1 & 1 \\ 1 & 0 & b & b \\ 1 & b & 0 & b \\ 1 & b & b & 0 \end{vmatrix}$$

f)
$$\begin{vmatrix} x & 0 & 0 & y \\ y & x & 0 & 0 \\ 0 & y & x & 0 \\ 0 & 0 & y & x \end{vmatrix}$$

3.3 Summation

1. Note that $\sum_{k=1}^{n} k^0 = n$ and $\sum_{k=1}^{n} k^1 = n^2/2 + n/2$.

Guess a relationship between the highest degree term of $\sum_{k=1}^{n} k^m$

and $\int n^m \, dn$, then prove this relationship if you can. Use your computer algebra systems to experimentally determine all of the

terms in $\sum_{k=1}^{n} k^m$ for several successive m beginning with m = 2,

and use the system to inductively prove each of your formulas. Then see if you can devise a formula or an algorithm that works for arbitrary nonnegative integer m.

3.4 Generating Functions and Power Series:

1. Using your computer algebra system, verify the following power series and determine their intervals of convergence:

$$(1 - x)^{-1} = 1 + x + x^2 + x^3 + \ldots$$

$$(1 - x)^{-2} = 1 + 2x + 3x^2 + 4x^3 + \ldots$$

$$(1 + x)/(1 + x + x^2) = 1 - x^2 + x^3 - x^5 + x^6 - x^8 + x^9 - \ldots$$

$$(1 + x)/(1 - x)^3 = 1 + 4x + 9x^2 + 16x^3 + 25x^4 + \ldots$$

Then, using these as building blocks or inspirations, see if you can experimentally discover rational expressions having the following power series expansions:

a) $1 - x + x^2 - x^3 + x^4 - x^5 + \ldots$

b) $1 + 2x + 4x^2 + 8x^3 + 16x^4 + 32x^5 + \ldots$

c) $1 + 2x + 3x^2 + x^3 + 2x^4 + 3x^5 + x^6 + 2x^7 + 3x^8 + \ldots$

d) $1 + 2x + 3x^2 + 2x^3 + x^4 + 2x^5 + 3x^6 + 2x^7 + x^8 + \ldots$

3.5 Integration and Differentiation

1.   Use your computer algebra system to evaluate the indefinite integral of $x^n e^{a\,x}$ for increasing n beginning with n = 0, until you can infer the general form. Then use the system to help you inductively prove that form.

2.   The size of successive partial derivatives can grow rapidly, especially if the original expression involves nested function compositions or nontrivial denominators. Find a particularly compact and innocent looking expression whose successive derivatives grow remarkably. The most dramatic example earns a prize!

3.6 Non-linear equations

Symbolic systems can also be used in an exploratory manner with more advanced topics. An example of this is given in Ollongren (Supporting Papers). Starting from a basic result in analysis (the generalized Taylor expansion), it is shown how to derive a general procedure for determining symbolic solutions of non-linear systems of equations containing a small parameter. The process is oriented towards computer algebra and is kept close to the formal theory. It is therefore transparent and easily applicable, although it may be an inefficient method for higher-order solutions. This latter shortcoming may, in fact, be readily remedied. The main point, however, is that modern computer algebra is now an important working tool in the hands of the practising mathematician.

References

Hosack, J., et al: 1983, A Proposal for a New Curriculum, Proposal to the Alfred P. Sloan Foundation.

Small, D., et al: 1983, Calculus of One and Several Variables: an Integrated Approach, Colby College.

Stoutemyer, D.R.: 'A Radical Proposal for Computer Algebra in Education', ACM SIGSAM Bulletin.

COMPUTER AWARE CURRICULA: IDEAS AND REALISATION

Hugh Burkhardt
Shell Centre for Mathematical Education,
University of Nottingham, England.

1 INTRODUCTION

Before getting down to the task of throwing ideas and comments into the pool which this meeting provides, there are some general points to be made about the nature of the exercise. It is speculative – a conference for conjectures; as in mathematics itself such activity is creative and important, but the outcomes should be seen as entirely provisional. We can have no reliable idea how far any suggestions we put forward will prove feasible in any, let alone every educational system. Even if they are implemented reasonably faithfully, the full curriculum reality of what occurs will contain many surprising side effects; more likely, the translation from an idea to a small scale pilot experiment with exceptional teachers and facilities, and then to large scale reality will involve critical distortions of the aims of the exercise which may call in question its value.

In case there are any who believe that I exaggerate the dangers, let me draw attention to a few examples so everyone can see what I have in mind:

The splendid Bourbaki enterprise was launched to establish a firmer foundation for undergraduate and graduate mathematical education; few now see that as among the positive contributions it has made, while many are concerned at the over-emphasis on formalism that has widely emerged from the movement.

SMALLTALK was devised by the Xerox Learning Research Group largely to produce a medium, the DYNABOOK, that would be "as natural to a child as pencil and paper" (1); what has emerged is perhaps the most sophisticated graphics orientated data management system so far – an important achievement, but a very different thing. SMALLTALK has not, at any rate, done any harm to the school curriculum, and its offspring, such as the Mackintosh microcomputer, may yet contribute.

My final example must be the reform movement of 25 years ago in mathematical education – "new math", "modern

mathematics" and so on. Comparison of the initial aims
agreed at conferences such as this, the pilot schemes in
a few exceptional schools, and the classroom reality of
today show the contrasts vividly. For example, in
England the applications of mathematics occupied a
central place in the original design; in most of the
major courses that emerged applications are mentioned
only to illustrate techniques with no serious attention
to the practical situations involved. Equally, new
mathematical concepts were introduced but often with
none of the pay off that motivated their inclusion –
because the serious examples originally envisaged
proved too difficult for most students, and were
replaced with trivial ones.

What are we to do about this? This is not the place for a serious
discussion of methodologies of research and curriculum development (2).
Very briefly, there is no proven successful answer but some seem to
be less susceptible to such corruption than others. I believe that
the essence is an empirical approach – find out what actually happens
to your draft ideas in practice, in circumstances sufficiently
representative of what you are aiming for, and then revise the
materials repeatedly until they work in the way intended. We have
found that structured classroom observation is a key ingredient in
our approach (3).

One other unusual factor makes curriculum development involving
advanced technology more difficult than usual. It is the mismatch
of time scales between technical change (one year) and curriculum
change (ten years). The curriculum designer can not assume a
specific level of technological provision and sophistication in
schools – it will vary widely both in time and from place to place.

This is important. If each student has a "micro", curriculum
possibilities open up which are not there with one micro per class;
these possibilities depend on the sophistication of the micro – one
line of display, a few lines, many lines, graphics, access to data –
each step is highly significant. Equally it is already clear that
low levels of provision and sophistication still have enormous
educational potential. Is technical restraint a virtue, or does it
impede progress?

The other targets that should be thought about are the teachers and
the students. I shall say little about the latter because they will
not be forgotten. It is the teachers that will face the greatest
difficulties; changing well established ways of working is extremely
difficult, particularly when teaching style is involved – as it must
be. As in any other highly skilled occupation, levels of performance
of mathematics teachers vary enormously; what works for the exceptional
few will not usually be accessible to a broader target group. The
situation is very different in the secondary (16-19) and tertiary

(18-22) centres. In many countries the greater independence of the
tertiary teacher, who has more control over curriculum and assessment,
means that it is easier to make experimental reforms but harder to
implement them on a larger scale; the stricter curriculum constraints
on the secondary teacher place heavy responsibilities on the innovator
who aims at large scale change.

It is not enough to talk of in-service training as a solution to such
difficulties; it has to be shown that it will be effective. Studies
and history suggest that changes of syllabus content have been
achieved but that, except for a small minority of teachers, changing
the pattern of classroom learning activities has not so far proved
possible.

This, however, is widely regarded as the central challenge of
mathematical education. Everywhere the curriculum is dominated by (4)

   Teacher explanation + illustrative examples + imitative exercises.

This leads to more rapid apparent student progress, but the skills
acquired are not usable on non-routine problems or in the world
outside the classroom. To achieve the flexible competence of understanding that this requires, the pattern of classroom activities has
to be widened to include some which give more initiative to the student.
It is encouraging that the micro has shown great promise in this
regard.

All change is threatening. Technology appears to reduce this threat,
partly because it produces an obviously new situation and thus does not
imply criticism of the teachers' existing modes of operation. This more
than compensates for the extra barrier of learning to use the equipment - provided it is reliable.

2   CHANGES IN MATHEMATICS
I shall not say much under this heading, because it has
received a lot of attention; what I say will relate fairly directly to
curriculum questions. The main areas of change in mathematics are
outlined in the background paper for this meeting (5). Many of the
issues are much more general than the technological background that
brought them to the focus of our attention. This is often so, and is
equally important in the curriculum and classroom dynamics domains.

Decisions on how far any change penetrates at any level will only
emerge from experiment. There are interesting questions for research
here. Their relevance to mathematical education will be slight unless
that is their focus. Forefront developments in mathematics or any
other subject do not often impinge on the taught curriculum - "modern
mathematics" in schools was almost entirely 19th century, and the same
is broadly true of undergraduate courses. Recent developments will
have to justify a curriculum slot against stiff competition, as well
as entrenched opposition; we need the evidence to support their

inclusion in curriculum terms. Ideas for such studies would be a
useful outcome of this meeting.

It seems to me that the central challenge of any new medium is to
acquire enough skill with, and understanding of it, so that it
becomes a powerful tool - and not a nett-absorber of effort and
attention. Otherwise, one is replacing mathematics by computer
studies - another possibility but not our goal here. The ambition
to provide a resource as natural to the mathematician, and to the
mathematics student, as pencil and paper, remains a good one. It
will not be easy. We shall be able to provide procedures for the
student to <u>follow</u>, as at present, which may well bring some further
insight - the danger is that we shall be content just to provide
more of this kind of fairly passive, imitative learning.

Thus, if the new medium is treated seriously, it will probably bring
better understanding but take more time. It should bring about a
<u>reduction</u> in total syllabus content. For example, the interrelation
between numerical, graphical and analytic methods of handling a
mathematical situation, their respective strengths and weaknesses, is
not easy to master but is essential both for understanding and for
action. The normal path of curriculum development, for example
the movement to introduce more "discrete mathematics", is likely to
lead to the opposite effect. The present course on calculus will not
prove dispensable and history suggests that the tendency will be to
arrive at a compromise with greater total content than at present;
this inevitably leads to an even greater emphasis on imitation. The
alternative of earlier specialisation avoids such hard choices by
transferring them to the student.

In one area there seems likely to be clear gain. The new central role
of algorithms, including their design rather than simply their
execution, is a rich field for developing both technical and higher
level skills. Algorithms are, I believe, inherently less abstract
that the implicit relationships (such as equations to solve) that
dominate the mathematics curriculum. The work of David Johnson and
others at Minnesota in the 1960's and 1970's showed that programming
could provide a semi-concrete bridge to abstract thinking that
enabled many more children to achieve some fluency in school algebra.
It is likely that similar gains can be established in the 16-22 age
range. It may be useful to take a broader, less formal view of
algorithms, with emphasis on graphical processes. Perhaps even human
processes, such as negotiations of criteria in solving a problem, may
usefully be brought within the algorithm framework.

Finally, another word of warning - because of the imitative nature of
the curriculum, it is easy to get a quite false picture of the <u>student
as mathematician</u>. A mathematician has command of a range of concepts
and techniques (or knows where and how to get such command) and uses
them autonomously to express and manipulate ideas and relationships to
get answers and understanding. There is clear evidence that, on such

criteria, students are <u>several years at least behind</u> their performance
on imitative exercises. The <u>calculator</u> is a useful resource because
students can use arithmetic for a range of purposes; in contrast it
has been shown (6) for example, that even very bright 17 year old
students may not use algebra at all as an autonomous mode of expression,
though they have had 5 years of success in manipulating it; the benefits
of a machine that will manipulate in a language they do not speak are
elusive, and maybe illusory.

3   CURRICULA

Computers and informatics can influence the mathematics
curriculum in at least two different ways. Some new developments in
mathematics will displace part of the current <u>content</u> because we come
to believe that students should learn about them; I shall not say much
more about such content aspects, which attract more attention than the
development in the student of the fundamental <u>processes</u> of doing
mathematics.

However, it seems to me that exemplary teaching "packages" rather than
general ideas on content will be needed both to convince and to enable
(7). We have begun to make some progress beyond speculation. Computing options are popular in undergraduates courses in mathematics, at
least in Britain, though they are rarely well integrated with the rest
of the mathematics curriculum. It will be most interesting to see the
results of the 20 experimental US college courses in discrete
mathematics funded by the Sloan Foundation, particularly when some of
them are developed and trialled by more representative teachers than
the initial innovators. It is worth keeping in mind the typical text
book for college calculus courses which stands as an exemplar and a
warning of what lies in wait at the end of the road of routine
development.

In other cases, there is such clear opportunity for the computer to
play a role (an introductory course on differential equations is one
obvious example) that it seems scandalous that courses have been
taught without – until we appreciate the difficulties of curriculum
change. There are many such developments of current courses to be
pursued, and surely collaboration, or at least communication, could
help.

At least as important as new content are the insights and opportunities
that computers provide in helping us tackle more effectively some of
the key problems in the mathematics curriculum; these are centred on
mathematical processes, particularly related to the development of
higher level skills. There is already some evidence that these
possibilities are rich and various; it is equally clear that we are
only at the beginning of discovering what they are.

Many of them need not have involved the computer. For example, it
happens that <u>mastery</u> is often expected in programming (you go on until
it works) but <u>rarely</u> in other parts of the mathematical curriculum.

Yet the mastery of a technique is essential if it is to be used in problem solving, pure or applied. Similarly, debugging skills are recognised as an essential element of computing. Research suggests that they are equally important in the mastery of mathematical techniques - effective mathematicians "debug" their half-remembered algorithms. The "diagnostic teaching" approach is designed to build on this.

The microcomputer has been shown (8) to be a powerful support to teachers in widening their style range to support more open activities; the design of programs to this end is an important field. The teaching skills involved then seem to transfer, at least to some extent to other teaching. It may well be at this stage, this is the most valuable single area for development - it is of course, a form of INSET as well.

The background paper rightly emphasises the curriculum opportunities for exploration, for "experimental mathematics", that the computer provides. However, we have a lot of evidence and some understanding of how difficult such activities are for the teacher to handle in the classroom.

Exploratory investigation as a key element in the curriculum has been a major objective in English mathematical education for at least 30 years - the Association of Teachers of Mathematics was founded largely to promote it. Despite strenuous efforts it has not happened except in a tiny minority (much less than 1 percent) of classrooms. Though the computer can provide support to teachers in this regard, the development of an investigative element in the curriculum can succeed only if it confronts the difficulty such activities present, particularly for teachers.

Equally, the challenge to explore must be at a level matched to the student - if the aim is to "discover" in an hour or so some important mathematical achievement that took a genius half-a-life-time to create, the exploration will have to be so closely guided as to be essentially a fake; on the other hand, interesting, though less global problems do exist at every level which the student can tackle on his own resources. For example, programming projects, at school and university have shown the possibilities and the difficulties for the teacher; a creative and systematic program of detailed empirical development will be essential if exploration is not to degenerate in most classrooms into that closely guided "discovery learning", which is really an alternative style of explanation.

We already have evidence (9,10) that the potential of the microcomputer for helping teachers to enhance student learning presents a tremendous opportunity for curriculum enhancement. The effects on the dynamics of the classroom can be profound, but they are often subtle; for this reason there is a great deal still to do before we have even a broad understanding of what can happen in the various modes of computer use of the kind listed in the background paper.

I shall illustrate the sort of thing that may be expected by describing
one application that has been developed and studied in some detail, and
which has proved particularly rich - the use by the teacher of a single
micro in the classroom programmed to be a "teaching assistant". I do
so for various reasons - it is less familiar to most people, it brings
out some general points about the overwhelming importance of the
people, teacher and pupils, and of the dynamics of their interaction,
and it is particularly relevant to schools as we know them because it
seeks to enhance the performance of a teacher working with a group of
children in the classroom in the normal way. It also only requires one
micro computer per class rather than one per child.

This mode of use, first emphasised by Rosemary Fraser, has been shown
to have remarkable effects in leading typical teachers in a quite
unforced and natural way to broaden their teaching style to include the
"open" elements that are essential for teaching problem solving. Since
this is a crucial aim that we have been trying to achieve for at least
thirty years with little or no effect, this is a valuable result. It
is worth explaining briefly why these effects come about (8). First,
the micro is viewed by the students as an independent "personality".
It takes over for a time a substantial part of the teacher's normal
"load" of explaining, managing, and task setting. These are key roles
played by every mathematics teacher. The micro takes them over in such
a way that the teacher is led into less directive roles, including
crucial discussion with the children on how they are tackling the
problem, providing guidance only of a general strategic kind -
counselling if you like.

It is equally important to recognise that there will be disappoint-
ments - or at least frustrations.

Apart from programming itself, perhaps the first big idea for using
computers in mathematical education was in teaching technical skills,
particularly arithmetic. The approach followed the behaviourist
teaching machine model. This has proved a much harder problem than
was expected. It is still unsolved. It seems that the computer can
be effective in teaching facts and straightforward techniques to
people who have little difficulty with them; so, of course, are other
methods. However, despite great efforts by some extremely talented
people, it has not so far proved possible to write programs which are
successful in diagnosing and remediating students' errors in technical
skills that they find difficult.

In other cases, the size of the potential "target group" is unclear.
The activities of that small proportion of enthusiastic "computer nuts"
display a motivation and the deployment of a range of strategic and
technical skills that are rarely matched in the normal curriculum.
(Could we ever visualise a mathematical "hacker" causing ingenious
chaos in the school or college mathematics department?) How far can

such rich learning activities be stimulated and exploited in all children? We do not know, but the proportion of children who are spontaneously using their home computers in this sort of way does not seem to be large. Again experiment is needed; we are hoping to set up such a study of "100-micro schools".

In tertiary education, it is common for the teacher to play a much narrower range of roles - explaining and task setting, with little else. The enrichment of the range of learning activities through the alteration of the classroom dynamics which the computer makes possible may not be welcome here; it makes a much more serious departure from standard lecture format than in schools (some at least) and, again, will certainly slow down the rush from one topic to the next which ensures a syllabus content of "high standard", whatever the level of independent student performance. This is particularly serious in advanced undergraduate pure mathematics courses, where enough 1 hour advanced problems often seem hard to find.

The questions I have raised require a great deal of work, integrating research techniques with curriculum development, before we have even a basic understanding of the classroom potential that we see. Experience suggests we shall find other possibilities of at least as much promise.

In order to realise the potential of any of these possibilities they will need to be systematically developed in detail with representative samples of teachers and students, using structured detailed data from the classroom.

REFERENCES

1 Goldberg A. (1978). 'Informatics and Mathematics in the Secondary School', 1977 IFIP Conference, Ed. D. Johnson and D. Tinsley, North Holland.
2 See, for example, Burkhardt H. (1982). 'How might we move the curriculum', 1982 BSPLM Oxford Conference, Shell Centre.
3 Burkhardt H. et al (1982). Design and Development of Programs as Teaching Material, Council for Educational Technology.
4 See, for example, Aspects of Secondary Education, Report of the HMI Secondary Survey, (HMSO, 1977).
5 ICMI, (1984). 'The Influence of Computers and Informatics on Mathematics and its Teaching'. L'Enseignement Mathématique 30, 159-172.
6 Treilibs V., Burkhardt H. and Low B. (1981). Formulation Processes in Mathematical Modelling, Shell Centre.
7 See, for example, 'Problems with Patterns and Numbers', a module of the Testing Strategic Skills programme (Joint Matriculation Board, Manchester M16 6EU, 1984).
8 Fraser R. et al (1983). Learning Activities and Classroom Roles, Shell Centre.

9 Burkhardt H. (1984). How Can Micros Help in Schools: Research
        Evidence, Shell Centre.
10 Burkhardt H. (to appear). 'The Microcomputer: Miracle or Menace in
        Mathematical Education' in Proceedings of ICME 5, Ed. M.
        Carss, Birkhauser.